BASIC BLACKSMITHING

BASIC

BLACKSMITHING

An introduction to toolmaking
with locally available materials

DAVID HARRIES and BERNHARD HEER

Practical Action Publishing Ltd
27a Albert Street, Rugby, CV21 2SG, Warwickshire, UK
www.practicalactionpublishing.org

ISBN 1 85339 195 6
ISBN 9781853391958

A catalogue record for this book is available from the British Library.

ince 1974, Practical Action Publishing has published and disseminated books and information in
support of international development work throughout the world. Practical Action Publishing is a
trading name of Practical Action Publishing Ltd (Company Reg. No. 1159018), the wholly owned
publishing company of Practical Action. Practical Action Publishing trades only in support of its
parent charity objectives and any profits are covenanted back to Practical Action (Charity Reg. No.
247257, Group VAT Registration No.880 9924 76).

Typeset by J & L Composition Ltd, Filey, North Yorkshire, UK

CONTENTS

GLOSSARY   vi

INTRODUCTION   vii

1. Tools and equipment   1
    A range of blacksmithing tools and equipment and their
    uses

2. Raw materials   5
    The uses of a range of raw materials commonly used
    by blacksmiths in rural areas

3. Blacksmithing techniques   7
    The basic techniques used by the blacksmith and
    information about forging temperatures and heat
    treatment

4. Making your own tools   23
    Step-by-step instructions on how to make the
    following tools:
    ○ Round punch   23
    ○ Hot chisel   25
    ○ Cold chisel   27
    ○ Hot and cold sets   31
    ○ Tongs   41
    ○ Fullers   47
    ○ Hammers   48

5. Products   75
    Step-by-step instructions on the following:
    ○ Axe-making   75
    ○ Hoe-making   82
    ○ Knife-making   91
    ○ The sickle   95

6. Setting up a workshop   101
    Hints for those who wish to set up their own workshop,
    including how to make bellows, hearths and anvils

APPENDIX: Forged tools for the carpenter   111
    Step-by-step instructions for making chisels, plane irons and a
    carpenter's brace and bit

QUESTIONNAIRE   127

GLOSSARY

Anti-roll bar	Carbon steel bar linking together the left and right front suspension units on some cars and vans.
Bevel	The sloping surface just behind the cutting edge of a bladed tool.
Bick	The round tapered part of the anvil.
Carbon steel	Steel with a high enough carbon content to be hardened by heat treatment.
Chamfer	A corner with the sharp edge removed, or the action of removing the sharp edge from a corner.
Clinker	Impurities from the fuel which gather in the bottom of the fire. If not removed these will block the air draught and stick to the metal as it is heated up.
Cross-section	A view of something, as if it had been cut through with a knife.
Fire-welding	Welding together pieces of metal using only the hammer and the heat from the forge. This is often called forge-welding.
Flux	Sand or other substance used to help ensure a clean join when fire-welding.
Fuller	Tool similar to a chisel or set but with a rounded end instead of a blade.
Fullering	Making a groove in a piece of steel with a fuller.
Hardie hole	Square hole in an anvil in which tools can be held.
Pritchel hole	Round hole in an anvil used when punching holes.
Quench	Place the metal in cold water or oil to cool it.
Reins	The long handles on tongs.
Scarf	The end of a piece of metal which has been prepared for a fire-weld.
Splines	The grooves which are often found cut into the ends of half-shafts.
Tang	The pointed part of a tool onto which the handle is fitted.
Temper	Heat treatment used to reduce the brittleness caused by hardening.

INTRODUCTION

This book explains basic blacksmithing techniques and gives step-by-step instructions on how to make a range of tools and products. It is divided into six sections, each of which has been designed to follow on from the one before. By working through the book in the order in which it is written, you can develop your skills while at the same time building up a set of tools. Each stage is illustrated by a drawing. There is also a section covering workshop equipment, including designs for bellows, hearths and anvils. (It is assumed that you will learn and practise blacksmithing techniques at an existing forge before committing yourself to setting up a workshop of your own.)

Starting with only an anvil, a pair of bellows and a few basic tools, almost all the tools needed by a blacksmith can be made from commonly found materials. Where possible, more than one method of making an item is described and more than one source of metal is suggested.

The tools in this book are not intended to copy those made in factories, but they work well, they cost little and they can all be maintained in the rural forge.

The designs of all the items are based upon the experiences of the authors while working with rural blacksmiths in Zimbabwe and Malawi. The book does not reject traditional techniques or equipment, but seeks to offer viable choices to enable smiths to develop new skills should they wish to do so. Where traditional solutions may be appropriate, for example goat-skin bellows in Africa, these have been included.

Throughout the book the authors have tried to avoid using fixed dimensions. The exact size and shape of a piece of metal may be important to an engineer making a part for a machine, but for the blacksmith these will vary according to the wishes of the customer and the metal in the scrap pile. The methods described are not the only ones that may be appropriate, but they have been chosen because they are accessible to most people.

Blacksmith at work in traditional African forge

1. TOOLS AND EQUIPMENT

Below is a list of basic tools and equipment appropriate to the needs of the rural blacksmith with a brief description of their use. Chapter 6 gives more information about this equipment.

Bellows

Used to drive air into the fire so that the fuel will burn at the high temperatures needed for forging steel. Four different types of bellows are shown in Chapter 6.

Hearth

The place where the blacksmith's fire is made.

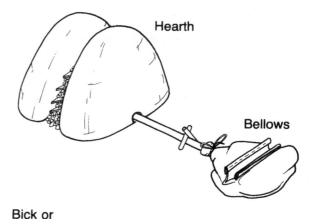

Anvil

A block of hard metal (or stone) on which the blacksmith hammers metal into shape.

Forge

The work area of the blacksmith. This word can be used to describe the whole workshop or just the hearth and bellows.

Sledgehammer

A heavy hammer, normally used two-handed by the blacksmith's assistant.

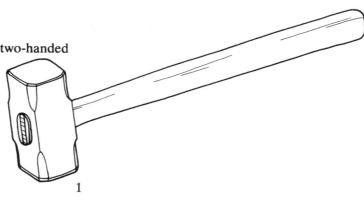

1

Cross-peen hammer

The hammer used by the blacksmith for most forging operations. It has one flat face and a peen as illustrated.

Round punch

Used instead of a drill to make holes in metal.

Hot chisel

Used to cut hot metal. This tool is not hardened in any way and should never be used to cut cold metal.

Cold chisel

Used to cut cold mild steel. This tool is hardened and tempered and should never be used on hot metal.

Hot set

Type of hot chisel with long handle, usually struck with a sledgehammer. This tool is not hardened in any way and should never be used to cut cold metal.

Cold set

Type of cold chisel with a long handle for heavy cutting. This tool is hardened and tempered and should never be used on hot metal. The cold set looks the same as the hot set except that the blade is considerably thicker and is ground to a different angle.

Cold set

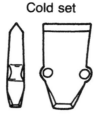

Tongs

Used to hold hot metal while it is being worked upon.

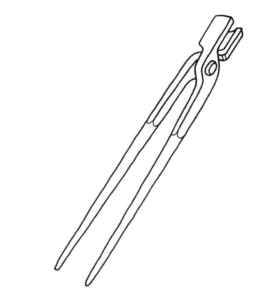

Top fuller

Used to make grooves in hot metal. A round-section piece of mild steel bar can be used as a simple top fuller.

Bottom fuller

Normally used with a top fuller, the bottom fuller rests on the anvil and is used to form a groove on the underside of the hot metal.

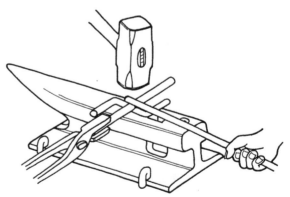

File

Used to clean up and sharpen many of the products of the blacksmith. Old files should be kept and used for hot-filing (see page 20).

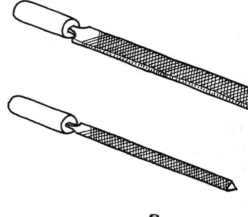

Water trough

Used for quenching metal during the hardening and tempering process. For most jobs a bucket can be used.

Other useful items

In addition to the tools and equipment listed above the following items will be very useful in the forge.

Empty cans

Used when tempering chisels or cooling down part of a piece of hot metal.

Piece of old grinding stone

Used to clean up hardened metal be tempering.

Clean sand

Used as a flux when welding.

2. RAW MATERIALS

The main sources of raw materials for these tools are scrap vehicle parts or scrap from construction sites.

Mild steel

The following are common sources of mild steel:

Sheet   Car body panel
Truck chassis
Oil drum

Bar   Reinforcing rods from construction sites (these often have a slightly higher carbon content than true mild steel)

Uses for mild steel include:

Sheet   Hoes
Hinges

Bar   Tongs
Set handles
Sickles (toothed pattern)

Working properties

Mild steel should be forged at a bright yellow heat as seen in the shade (just below the temperature where it starts to give off sparks). Some minor bending and cutting can be done cold. Mild steel cannot be hardened enough to use for most cutting tools. It is easier to fire-weld than other steels.

Medium-carbon steel

Many vehicle and machine parts are made from medium-carbon steel, including:

Vehicle half-shaft
Tractor plough disc
Plough mould-board

Uses for medium-carbon steel include:

Hammers
Hot chisels
Hoes
Punches

Working properties

Medium-carbon steels should be forged a little cooler than mild steel, but still at a yellow heat. The extra carbon in the metal makes it possible to harden it to some extent. With practice it can be fire-welded.

High-carbon steel

The following are likely sources of high-carbon steel:

Suspension coil spring
Torsion bar (used on the front suspension of some cars)
Vehicle leaf spring
Anti-roll bar (a sprung bar running between the two front suspension arms on some cars)
Wood saw
Large hacksaw blade

Uses include:

Cold chisels
Cold sets
Knives
Wood chisels
Plane blades
Axes
Adzes
Wood- and stone-carving tools
Tinsnips

Working properties

High-carbon steel is the most useful of all the steels to the blacksmith. All tools and implements which require a durable cutting edge should be made of high-carbon steel. It should be forged at an orange/yellow heat but not quite as hot as the medium-carbon steel. It can be hardened. It does not fire-weld easily but can with practice be welded onto steels with a low carbon content.

Fuels

The most common type of fuel for the rural blacksmith is charcoal. Charcoal is a very clean burning fuel; in other words, it produces very little clinker. Charcoal from hardwood trees has traditionally been the favourite fuel of many rural blacksmiths, so much so that in many areas all the hardwood trees have now been used up. Although hardwood charcoal is very good, softwood charcoal or charcoal made from oil palm husks can be used, provided it has been carefully made.

To get good quality charcoal, the wood ne to be burnt very slowly. A good way to do is to stack the wood in a pit and to cover it with earth while it is burning. This will re the amount of air that the fire can get and sh slow down the burning enough to produ fairly dense charcoal which will give out pl of heat when burnt on the forge. A char fire tends to spread and needs careful management in order to avoid wastage.

One other fuel which is sometimes availab urban areas is coke. Coke is a dense fuel w gives a very hot fire. It needs a strong air t to burn and will go out whenever the bellow not in use. It is more difficult to light than char and has impurities in it which will very qui form into clinker in the bottom of the fire

3. BLACKSMITHING TECHNIQUES

Fire management

Looking after the fire is an essential part of the blacksmith's work. It is important when lighting the fire to make sure that the fuel where the air blast enters is properly alight. As the fuel burns away clinker will build up just below the nozzle. If the clinker is not removed, it will soon choke the fire and start to stick to the metal which is being heated up. The clinker can easily be taken out by allowing the fire to cool slightly (this makes the clinker harden into a lump) and lifting it out on the end of a poker or rake. The clinker should be removed completely from the hearth or it may get mixed up with the fresh fuel and end up back in the fire.

Try to keep your fire just the right size for the job you are doing. If the fire is too large it wastes fuel and is difficult to control. If it is too small it will be difficult to reach a good working temperature.

Forging temperatures

Being able to recognize the right temperature at which to work metal is one of the most important skills of the blacksmith. Different types of steel need to be worked at different temperatures; the rule is that the lower the carbon content the hotter the metal can be worked.

Only by looking at the colour and the brightness of the metal can you tell when the steel reaches its correct working temperature. The temperature should always be judged in the shade as it is very hard to see how hot the metal is if the sun is shining on it.

The hotter the metal, the easier it will be to forge; however, if overheated, the metal will start to burn away in the fire. Metal with a high carbon content will start to burn at a lower temperature than mild steel. Forging high-carbon steel at the wrong temperature causes internal stresses which will weaken the metal.

The following list shows the colour and brightness at which different forging processes should be carried out:

Dull red	Hardening or annealing high-carbon steels
Medium red	Hardening or annealing medium carbon steels; minor bends in carbon or mild steels
Bright red	Annealing mild steels
Orange/ yellow	Forging of high-carbon steels
Yellow	Forging of medium-carbon steels
Bright yellow	Forging of mild steel
Bright yellow metal which looks oily in the fire and throws off a few whitish sparks	Welding

Basic techniques

The skill of shaping the metal involves a number of processes. The main techniques are:

o bending
o drawing down
o cutting
o upsetting
o punching and drifting
o fire-welding

These techniques should be practised before you attempt to make products.

Bending

Minor bends in mild steel can be made when the steel is cold; however, most bends are easiest with the metal at its normal forging temperature.

If you are bending the metal over the edge of the anvil, keep to the rounded area or the surface of the metal will be marked. Always strike the metal slightly in front of where you want the bend to be. Avoid crushing the metal between the hammer and the corner of the anvil as this will draw it down rather than bend it. The metal tends to bend where it is hottest, so be sure to heat the bar in exactly the right place.

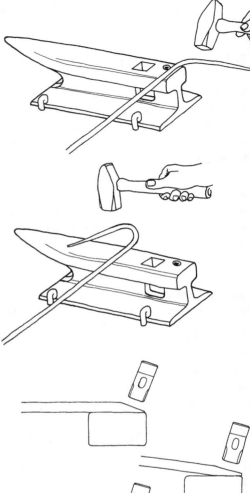

Drawing down

Reducing the thickness of steel by forging is called drawing down. Drawing down can be used to make a pointed tip on a piece of steel such as the tang of an axe. It can also be used to thin out a central part of a bar.

When you are drawing down, the angles between the hammer face, the metal and the anvil will decide the shape of finished work.

Stages in the process of drawing down are illustrated.

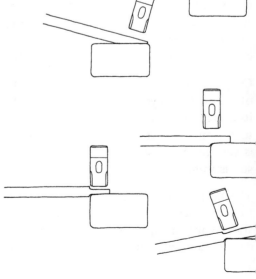

A flat chisel point

With the metal and the hammer held as shown, start to draw down the point. First, work the tip of the bar down to the right section. As you do this, the end will spread out like a fish's tail.

As soon as you see this starting to happen, lie the metal on its edge across the anvil and flatten the sides back in. Finally work back along the bar until you have the correct length of point.

A square point

Holding the metal and the hammer as shown, work first of all on the tip. Rotate the metal through a quarter of a turn every few blows until the tip is the correct size. Next, work back along the bar, still turning the metal every few blows until the taper reaches the correct length. If the tip is uneven it can be flattened with one or two light hammer blows.

A round point

Forge a square point as described above. Then, working from the back towards the tip, forge in the corners. The nearer to the tip you are working, the more gentle the hammer blows will need to be. Finally, remove the remaining ridges with gentle hammer blows.

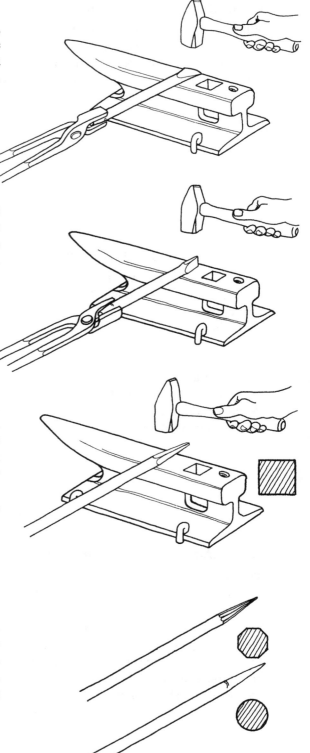

Drawing down parallel square

With the metal held flat across the anvil, forge the metal out, keeping the hammer face parallel to the top of the anvil. As with the square point, rotate the metal through a quarter of a turn every few blows.

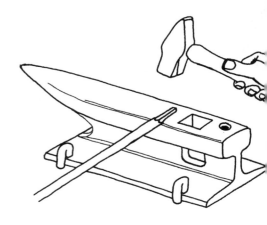

Drawing down parallel round

First of all, draw down to a parallel square section as described above. Next, remove the four corner ridges and, finally, round off with gentle hammer blows.

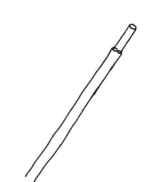

All of the above systems will work well when drawing down small section pieces of steel. When working on heavier pieces much time and effort can be saved if slightly different methods are used.

Using the peen of the hammer

The metal is drawn down first using evenly spaced blows with the peen of the hammer instead of the face. The marks left by the peen are then flattened using the face. The shape of the peen squeezes the metal out much faster than is possible with the face. This method also gives you control over the direction in which the metal is pushed, allowing you either to push the metal along the bar or to spread the metal out sideways.

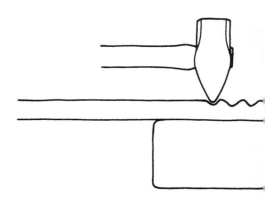

Using the bick of the anvil

The bick of the anvil can also be used to speed up and control the process of drawing down metal. The curve on the top of the bick acts in much the same way as the peen of the hammer. When a piece of steel is worked across the bick it will quickly lengthen without becoming much wider. If held along the bick it will spread out sideways without gaining much in length.

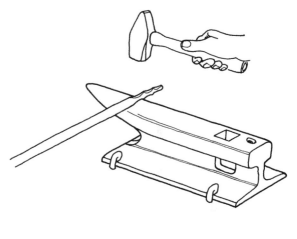

Using the rounded corner of the anvil

Using the rounded corner of the anvil will draw down metal very quickly. The bar to be drawn down is held at an angle and hammered exactly over the rounded corner of the anvil. Between each blow the bar must be moved along slightly to avoid making the metal too thin in one spot. As with the two previous methods, the dents made by the rounded surface must be removed using the face of the hammer and the flat part of the anvil.

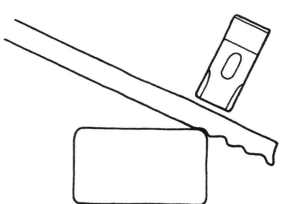

Cutting

Small section mild steel can be cut cold with either a cold chisel or a cold set. Carbon steels and large section mild steel should always be cut hot. This is done with either a hot chisel or a hot set.

Cold cutting

To cut small section mild steel with either a cold chisel or a cold set, the metal is first cut about a quarter of the way through from one side, then turned over and cut a quarter of the way through from the other side, and finally broken off by bending back and forth over the edge of the anvil. Blacksmiths who work only with spring steel will not find these tools useful. Cold cutting tools must never be used on hot metal as this will spoil their temper (see page 21).

Hot cutting

Cutting hot metal quickly and efficiently is for many blacksmiths their most important skill. Cutting is done using either a hot set or a hot chisel. Working with an assistant and using the hot set is the quickest method. With both tools the system of cutting follows the same rule an accurate cut is required the metal is bro up to a red heat and marked using the c or set.

The metal is then brought up to its normal working temperature and cut almost all the way through. Every few blows the blade of the cutting tool must be quenched or it will become damaged. The final blows should be either over the edge of the anvil or onto a mild steel cutting plate. Cutting the metal all the way through on the anvil face damages both the anvil and the tool.

When cutting out a quantity of items, time and fuel can be saved by only cutting halfway through each piece. The piece is then bent along the cut line and left to cool.

When the metal is cool, a few blows with the hammer will fracture it along the cut line. Hot-cutting tools should never be used to cut cold metal as they are not hardened. If used on cold metal the edge of the blade will be spoilt.

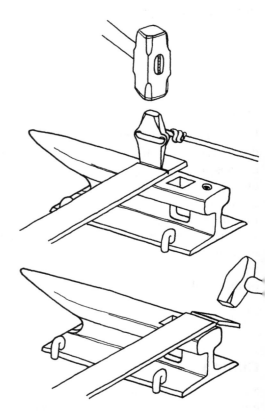

Upsetting

Increasing the thickness of a piece of metal is called upsetting or jumping up. During this process there is often a tendency for the metal to bend. Each time this happens, upsetting must stop until the bar has been straightened out. Bending is less likely if the parts of the bar which are not being upset are kept cool and if the tip of the bar is completely level before the process is begun. It can be done in several ways. Two of them are outlined below. In both cases the work should be done at near welding heat.

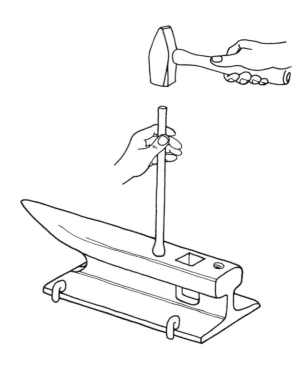

Method 1

Hold the metal vertically with the heated end resting on the anvil and strike the top of the metal as if trying to drive it into the anvil.

Method 2

This method is only suitable for large pieces of metal. Hold the metal vertically with the heated end just above either a steel block on the floor or the anvil. Bounce the metal on the anvil or block and the weight of the bar will force the heated end to swell out in all directions.

Sometimes it is necessary to upset a bar away from its end. To do this use the above methods but be sure to heat only the area to be upset. If you are working with mild steel, other parts of the bar can be cooled with water.

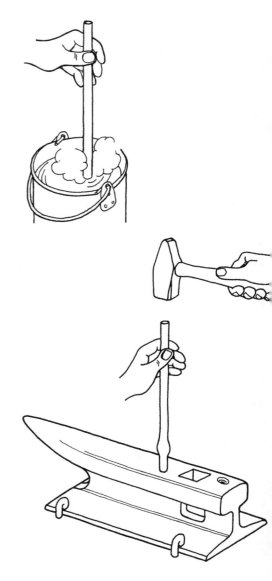

Punching

Punching is a very useful process; it allows the blacksmith to make holes of different shapes and sizes. Some of the metal from the hole is pushed into the surrounding area leaving the workpiece stronger than if the hole had been drilled.

To punch a hole, heat up the metal to the correct forging temperature. Drive the punch into the metal, working on the flat part of the anvil. Every few blows, cool the punch down to prevent it from overheating and losing shape. When the punch is nearly through, turn the metal over. The punch will have left a mark on this side.

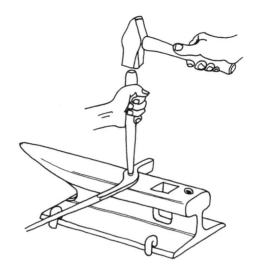

Quickly move the workpiece so that the mark is over the hole in the anvil and drive the punch through. The raised area around the hole can then be flattened back in.

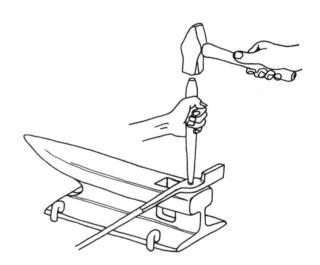

Fire-welding

Certain types of steel can be welded together using nothing other than the heat of the fire and the pressure of the hammer blow. The following welds can be successfully performed:

○ Mild steel to mild steel, for example making chain links.
○ High- or medium-carbon steel to mild steel, for example hard-facing a hammer.

Welding together two pieces of high-carbon steel is not practical in the forge.

A good weld can be produced if careful attention is given to the following:

The metal

Fire-welds require careful preparation. In all cases, care at this stage will help to ensure that a weld is successful.

The fire

The fire should be free from clinker and well covered with fresh fuel while the metal is brought up to welding heat.

Heating the metal

The pieces to be welded together must both reach welding heat at the same time. They should be placed in the fire a little distance above where the air enters. Large pieces should be heated up slowly to prevent the surface of the metal getting too hot before the inside has reached welding heat. The scarfs should face downwards to prevent clinker from settling on the parts of the metal which are to be welded.

Temperature

Recognizing the right moment to remove metal from the fire for a weld takes experie Too soon and the metal will not stick. Too and the metal may simply fall apart. At wel heat, apart from the brightness of the col there are usually two signs to look out for. of all, the metal will start to sweat. At this p the metal looks as though it is wet on surface. The other sign is that, as the met pulled slightly away from the heart of the small white sparks will start to jump from

Speed and technique

When the metal has reached welding hea time must be wasted between removing it the fire and performing the weld. The m should with one movement be brought o the fire, shaken clean (to remove any slag), brought into the correct position on the a The weld is then made using rapid blows, ge at first then fairly firmly once the metal stuck. As soon as the top scarf has been we in, turn the metal over and work in the bo scarf. The whole operation must be compl with the metal still at welding heat.

Performing a weld

If you have not welded before, the following exercise is well worth trying. Remember to start out with a clean fire.

○ Take a length of mild steel bar and flatten the end as shown.

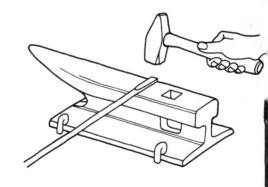

○ Shape the tip into a one-sided chisel point.

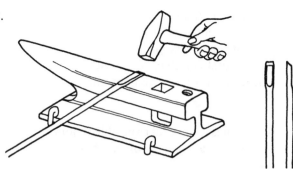

○ Cut about halfway through the metal in the middle of the flattened part.

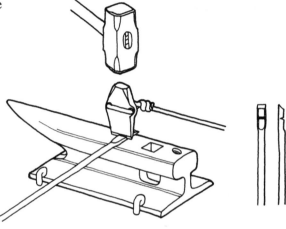

○ Fold the metal in half along the cut line.

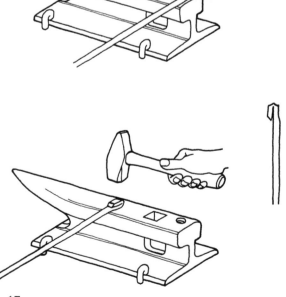

○ Put the metal in the fire with the folded area in the centre of the fire, a little above the point where the air enters. Make sure that the metal is well covered with fuel and bring it up to a welding temperature.

○ As soon as welding heat is reached, lay metal across the anvil and weld the fo▌ section together with rapid firm blows.

Sometimes it is useful to be able to join together two separate pieces of mild steel by fire-welding, using the following method.

First of all, scarfs must be formed on the pieces which are to be joined. Upset the ends of both bars.

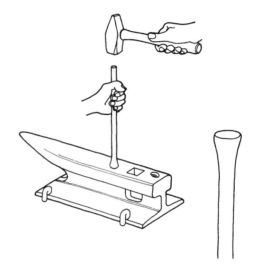

Forge in the tips until you have formed a short square section on the end of the bars.

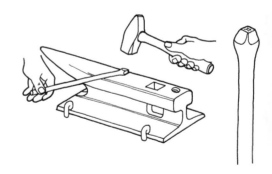

Using first the face of the hammer with the bar held level, and then the peen with the metal held at an angle as shown, forge the tip of the scarf.

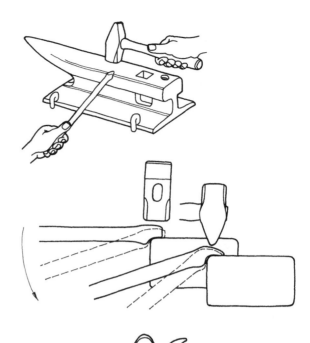

The finished scarfs should look like this when viewed from the top and the side.

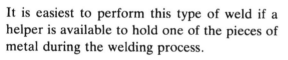

It is easiest to perform this type of weld if a helper is available to hold one of the pieces of metal during the welding process.

Place both pieces in the fire as described for the previous weld. The hollow section in the scarfs should always face downwards in the fire so that clinker does not form on the surfaces which are to be welded together.

Check that both pieces are heating evenly by removing them occasionally from the heat for inspection. If one appears to be cooler than the other, make sure when you put it back that it is the nearest of the two to the hottest part of the fire.

When both pieces have reached welding heat they should be brought out of the fire, shaken clean and laid across the anvil as shown.

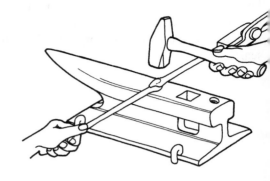

The weld should now be performed. The position of the first few blows on this type of weld is important and should be in the order shown here. This will help to ensure that all of the slag is driven out of the weld and that a good strong joint is formed.

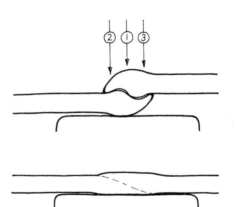

The finished weld should look like this.

Hot-filing

Hot-filing is a very useful technique for the blacksmith. To hot-file a piece of steel it is brought up to its normal working temperature and then filed to shape using an old file. The technique is especially useful when working with carbon steels which would normally have to be annealed before a file could be used on them.

An old file should be saved and only used for hot-filing because the heat from the metal will spoil the temper and make it too soft to use on cold steel.

This technique may be difficult without the use of a vice.

Heat treatment

There are four different types of heat treatment commonly used by the blacksmith, each of which has a different effect upon the metal. An understanding of all four will help you to produce better quality products and save time and effort.

The four processes are:

○ Annealing
○ Normalizing
○ Hardening
○ Tempering

Sometimes all of these processes will be used on a single job. Hardening and tempering is not necessary on tools which will be used on hot metal.

Annealing

Annealing is a process which softens carbon steels, allowing the metal to be filed, sawn or bent cold. It is also often used as a preparation for the hardening process as it is the most effective way of removing stresses from within a piece of steel.

Steel is annealed by slowly heating it up to red heat and then burying it in ashes. This slows down the cooling process. Small items are best buried alongside a heated piece of scrap steel to prevent them from cooling too fast. The longer the cooling period (up to around 10 hours) the softer the metal will become.

Normalizing

Normalizing is a process used to take out most of the stresses put into the steel during forging.

The metal should be heated to the same temperature as is used for annealing, and left to cool down naturally. The metal should not be left on the anvil during this process as it conducts the heat away too quickly and can cause hard spots.

Hardening

Medium- and high-carbon steels can be made hard but very brittle by rapid cooling. The brittleness can then be removed by tempering the metal after hardening.

The metal is heated up to a medium red heat and immediately quenched until cold. For most jobs the quenching is done in water. If the water does not cool the metal quickly enough to harden it, a little salt can be added. Occasionally the water may cool the metal down too quickly, causing cracks to form.

Quenching the metal in oil will slow down the cooling process slightly; if oil is not available, adding a little earth to the quenching water and mixing it to form a thin mud will make a good alternative.

Tempering

After the hardening process, the metal is very hard but brittle, in the same way that glass is hard and brittle. By re-heating the metal can be made tough again, but as the toughness increases the hardness decreases.

As the metal is heated up and these changes start to take place oxides build up on the surface of the steel. If the metal has been polished with a piece of old grinding stone these oxides can be seen, gradually changing colour as the metal

Temperature (degrees centigrade)		
Carpenter's chisel	230	Pale straw
Plane blade	230	Pale straw
Scraper	230	Pale straw
Hammer face	240	Straw
Flat drill	250	Dark straw
Scissor blade	260	Purple speckles
Cold chisel	270	Purple
Centre punch	270	Purple
Wood drill	270	Purple
Screwdriver	280	Dark blue
Spring	300	Pale blue

is heated. These tempering colours can be used as a guide, telling you how much toughness the metal has gained and how much hardness it has lost. On page 21 is a list of blacksmiths' products and the recommended colour to which they should be tempered. For all bladed items the tempering colour listed should be seen along the cutting edge of the blade.

Carbon steels do not all behave the same when hardened and tempered. If, after testing, the hardness of the tool appears to be wrong, heat treatment should be repeated using following as a guide:

○ If the finished tool dents when used (i.e. is too soft), anneal, re-harden and tempe a lower temperature.

○ If the finished tool chips when used (i.e. too hard), clean the metal and temper higher temperature.

4. MAKING YOUR OWN TOOLS

Round punch

Punches for most purposes can be made from used vehicle coil springs. If these are not available or if you want to make a very large punch, anti-roll bars or torsion springs can be used.

When finished the tool should look like this.

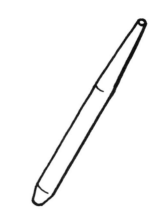

Draw a chalk line across your anvil about 200 mm from the end. Lay the spring across the anvil with its end level with the back of the anvil. Roll the spring along the anvil as shown.

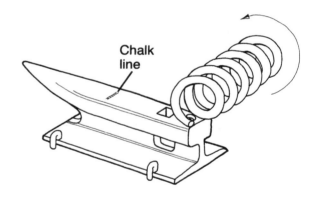

Chalk line

The part of the spring which is now level with the chalk line is the point at which the cut should be made. Put a chalk mark on the spring at this point and place it in the fire with the chalk mark in the hottest part.

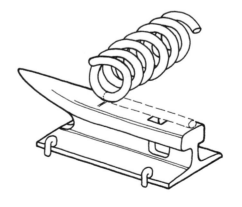

Cut almost all the way through the heated spring at the point where it was marked, using a hot set or a hot chisel. Break the piece off using a pair of tongs.

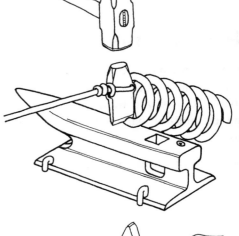

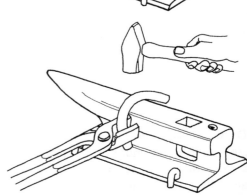

Straighten out the piece of coil spring as shown.

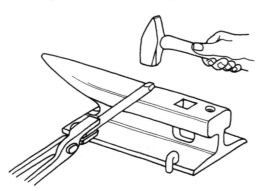

Next, form the striking end of the punch by drawing down and flattening back the end.

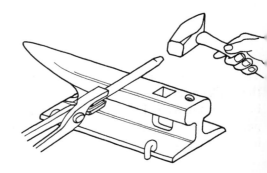

The workpiece should now look like this.

Carefully draw down the other end of the punch to a smooth, gentle round taper. Remember to start off by drawing down square (see page 9).

Put the finished punch to one side and allow it to normalize.

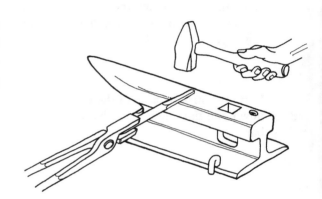

Hot chisel

The following instructions show how to make a hot chisel from an old car half-shaft. If this material is not available an anti-roll bar, torsion bar or any length of similar medium- to high-carbon steel can be used.

The finished hot chisel should look like this.

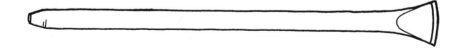

Using a hot set or a hot chisel, cut off the splined (grooved) section of the half-shaft. After every few blows, turn the bar slightly so that the metal is cut evenly from all sides.

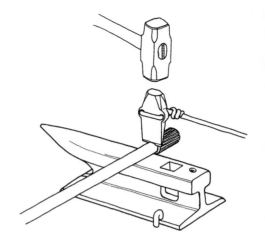

Remember to quench the blade of the cutting tool every few blows and to finish the cut over the edge of the anvil.

Form the striking area of the chisel by drawing down and flattening back the other end of the bar.

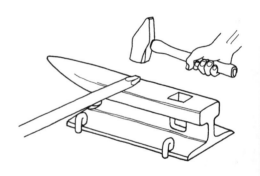

With a hot set or hot chisel cut the shaft to length. The body of the chisel being forged should be fairly long (about 400 mm). This allows the hand holding it to be kept at a safe distance from the hot metal when it is used.

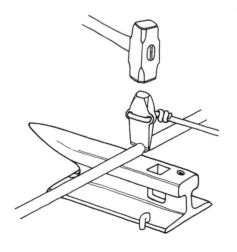

Draw down the blade section, allowing it to spread out. This will give you a long and slightly curved cutting edge.

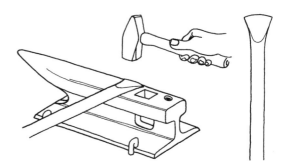

File or grind the cutting edge as shown.

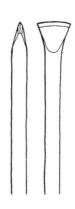

Cold chisel

A useful cold chisel can be made from a short length (about 150 mm) of vehicle coil spring. Apart from the length of the piece of steel required, the first three stages are exactly the same as for the round punch.

When finished the tool should look like this.

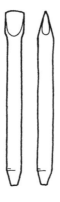

Using the system described on page 23, measure the length of the metal ready for cutting.

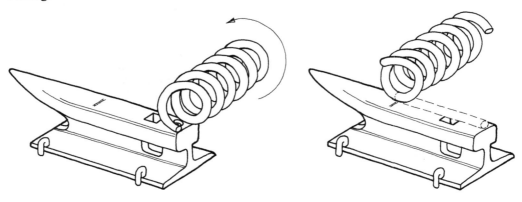

Place the spring in the fire with the chalk mark in the hottest part, until it reaches yellow heat.

Cut most of the way through the spring at the point where it was marked using a hot set or a hot chisel. Remember to quench the blade of the cutting tool every few blows. Break the piece off using a pair of tongs.

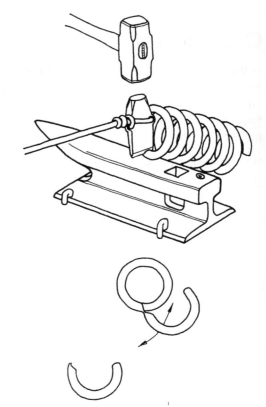

Holding the metal in a pair of tongs, and using the hardie hole to hold one end, straighten out the piece of coil spring and flatten back the ends as described on page 24.

Next form the striking end of the chisel by drawing down and once again flattening the end.

Carefully draw down the other end to form a flat-sided chisel point as described on page 9.

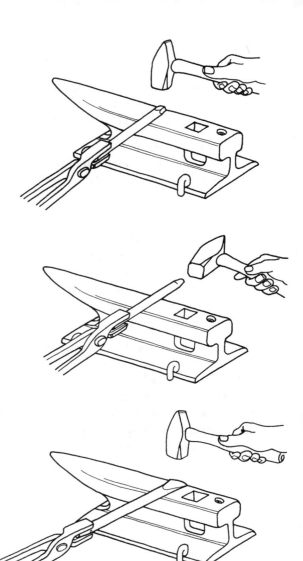

Hot-file or grind the chisel point as shown. Remember to use an old file to do the hot-filing.

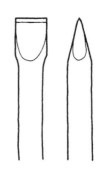

The final stage is the hardening and tempering of the chisel. Before attempting this be sure to read the section on heat treatment on pages 21–2.

Hardening and tempering a cold chisel should be done in one heat. You will need to have the following tools and equipment ready before you begin:

○ A bucket or container full of water.
○ A piece of an old grinding wheel for cleaning the steel after hardening. If this is not available a sandy or gritty piece of natural stone often works just as well.
○ A can or bucket with a little water just covering the bottom.

Bring the chisel point and about one-third of the body slowly up to a red heat. Be careful not to let the blade get too hot. Holding the chisel upright quench the tip of the chisel in the full container of water. Move the chisel up and down slightly while quenching without letting the tip come above the surface of the water.

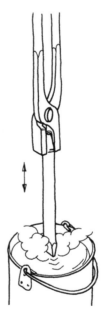

As soon as the tip has had time to cool down, remove the chisel from the water. Using the piece of grinding wheel, quickly clean the surface of the metal on one side of the chisel near to and along the blade. Avoid resting the chisel against the anvil while cleaning as this will cause it to cool down too quickly.

As soon as bright steel is showing, start to look for the temper colours. The heat which is still in the body of the chisel should now be spreading down towards the blade. With this heat will come a band of colours ranging from the colour of dry grass or straw through brown and purple to a pale blue. Just after the straw colour has reached the blade it should start to turn purple. This is the moment when the blade must be quenched in the second water container where it should be left with water just covering the blade to cool down.

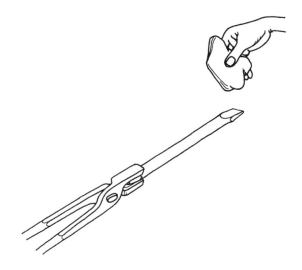

Hot and cold sets

The following section explains how to make hot and cold sets. A set is like a chisel with its blade attached to a long handle. This gives the tool two advantages over ordinary chisels. Firstly, the hand holding the tool is kept well away from the hot metal and secondly, when working with an assistant, the short but fairly heavy blade can be safely struck with a sledgehammer.

Hot set

The hot set illustrated can be made from a short length of fairly thick leaf spring and some 10 mm round section mild steel rod.

Using a hot set or hot chisel cut about halfway through a piece of leaf spring approximately 40 mm from the end. Remember to quench the cutting edge of the tool every few blows.

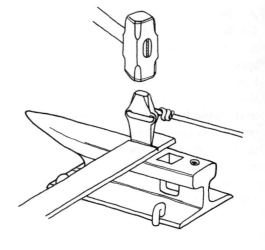

Break the metal along the cut line over the edge of the anvil.

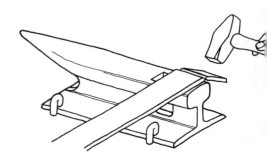

Holding the metal in a pair of tongs, forge in the rough edge where the metal was cut off. Hammering first on one side, then the other, forge the metal into a wedge shape as shown.

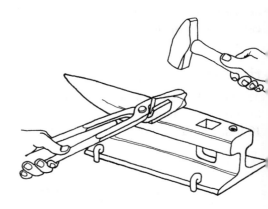

Flatten the top of the thicker end. This will form the part of the tool which is struck by the hammer when in use. Forge in the corners around this face.

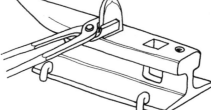

With the punch made earlier, punch about three-quarters of the way through the metal as shown.

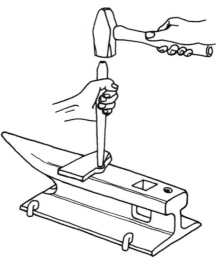

Turn the metal over and finish the hole off from the other side.

The hole should be opened out until it is just large enough for the mild steel rod to fit through.

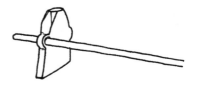

Repeat the process on the other side of the set. The workpiece should now look like this.

Forge down the blade section as shown.

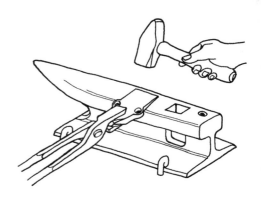

Work can now begin on the handle. Draw down the end of the piece of mild steel rod to a gently tapered round point.

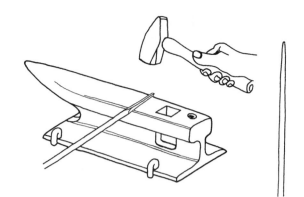

Approximately 150 mm from the end, bend the metal into a **U** shape. The width of the **U** should be the same as the distance between the holes punched in the tool head.

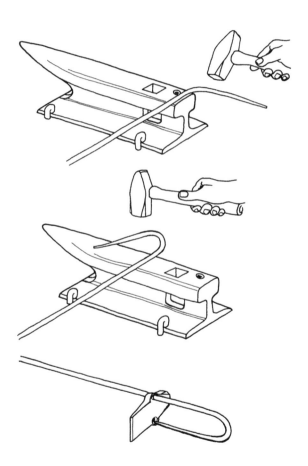

Slip the tool head onto the handle and flatten the **U** section as shown.

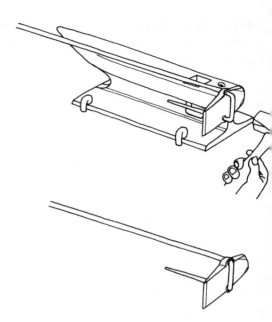

Fold the handle around the tool head so that the drawn-down section is left sticking out at one side.

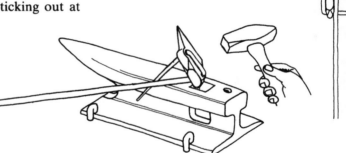

Holding the tool in a pair of tongs, as shown, wrap the drawn-down section around the bar. This will secure the tool head firmly to its handle.

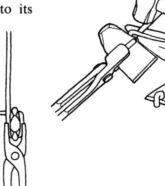

The workpiece should now look like this.

The other end of the handle can now be formed. Bend the bar through 90 degrees approximately 15 mm from the end.

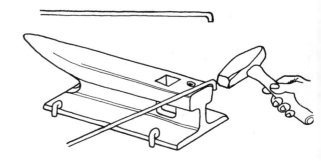

Finally, bend the bar over to form the handle, as shown in the drawing.

The blade can now be ground or filed to shape.

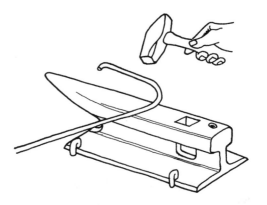

Wrap-around handles

With slight alterations to the techniques described above, handles can be made using thinner mild steel rod. The finished tool should look like this.

Instead of punching two holes in the set, a simple top and bottom fuller (as described on page 47) is used to form a groove as shown in the drawings.

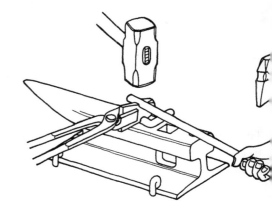

The tool head is then held in a vice or, if no vice is available, jammed in a log.

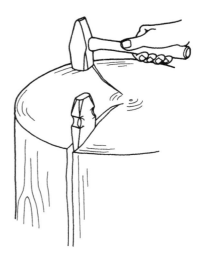

A length of mild steel rod is now bent into a **U** shape so that it will fit around the head of the tool. The rod should be a little over double the length of the handle you wish to make.

Very quickly, with the metal at correct working
temperature, wrap the rod around the head of
the tool as shown.

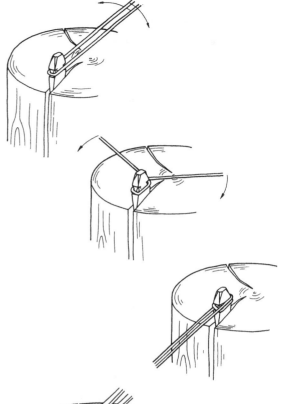

Using a pair of tongs, pull the set out of the
log or release it from the vice.

With a few gentle blows, forge the two rods
together at the point where they meet the head
of the tool.

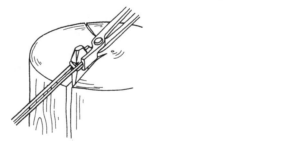

The tips of the rods can be either fire-welded
or twisted together using a pair of tongs.

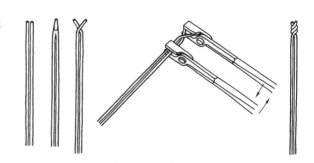

The two rods can now be opened out to form a comfortable handle. This is best done using the peen of the hammer over the bick of the anvil.

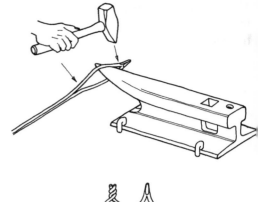

Finally, remove any sharp edges from the tip of the handle with a file.

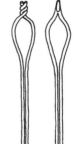

Tightening the handle

After some use, the handle may work loose. It can easily be tightened by putting a twist just behind where it is wrapped around the tool head. If necessary, this can be done several times before a new handle needs to be made.

Cold set

A useful cold set can be made using the same methods as for the hot set. The only difference is that the blade section should be left thick as shown in the drawing and the cutting edge should be hardened and tempered using the method explained on page 21. The illustration shows the difference between the blades of the hot and cold sets.

Hot set

Cold set

Tongs

This section shows how to make a pair of tongs from round mild steel bar. In the drawings two different sizes of metal are shown. Most of the work is forged from a piece 400 mm long by 20 mm diameter. A short length of 15 mm diameter is used for the rivet. This will make a pair of tongs suitable for light work such as making axes, adzes or hoes. Larger section steel can be used if tongs for heavier work are required. If mild steel bar is not available, off-cuts of steel reinforcing rod from construction sites can by used instead.

When finished the tongs should look like this.

Draw a chalk line along the anvil about 25 mm from the near edge. This line will help you to forge both jaws of the tongs to the same size. When you are striking the first few blows the end of the bar should be brought level with the chalk line. Keeping the bar pressed against the rounded part of the anvil draw down the jaw as shown in the picture. Flatten the sides and the end of the jaw.

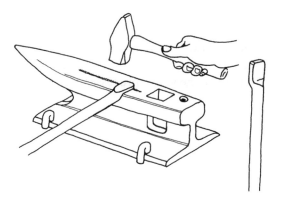

41

Rotate the bar a quarter-turn to the left and hold at an angle across the anvil as shown. Forge the flat area for the rivet.

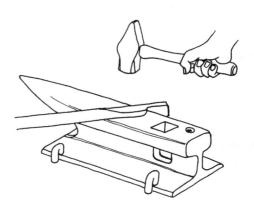

Rotate the bar a further quarter-turn to the left and start to draw down the reins (handles). Be careful to work only over the rounded edge of the anvil.

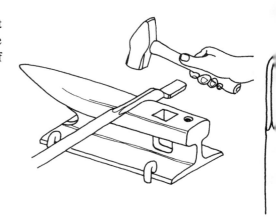

Draw down the rein as shown and leave the metal to cool down. Turn the bar round and repeat all the above processes on the other end.

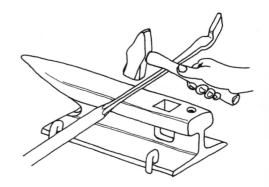

Cut the bar in the centre using a hot chisel or hot set.

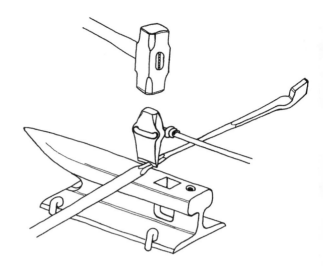

Finish drawing down the reins and round off the ends.

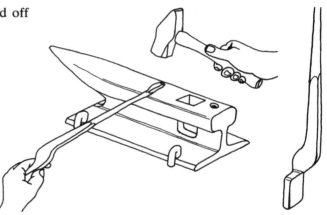

Working over the rounded part of the anvil (the bick), offset the jaws ready for punching as shown in the drawing.

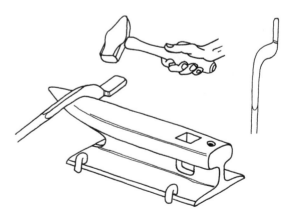

Using the round punch made earlier and the technique described on pages 33–4, punch the eye almost through, over the flat of the anvil with the jaw facing down.

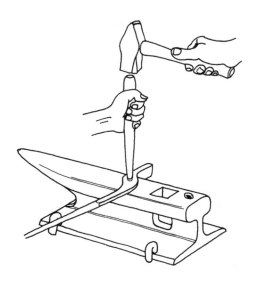

Turn the metal over and place over the pritchel (punching) hole. Punch the rest of the way through.

Before moving on to the next stage make sure that the two halves of the tongs are the same, and that they fit together properly.

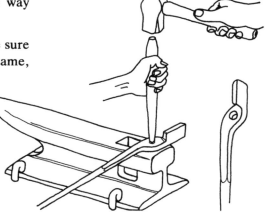

The next stage is to make and fit the rivet. Draw down parallel the end of a piece of round section mild steel bar until it will fit through the holes punched earlier in the two halves of the tongs.

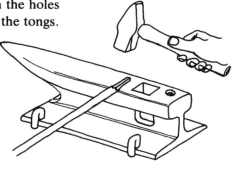

The drawn-down section should be just long enough to go through both halves of the tongs and leave twice its own thickness sticking out of the other side.

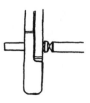

The rivet head should now be formed by cutting almost all the way through the bar, a little behind the shoulder.

Lay the two halves of the tongs together so that the holes are lined up with each other. Bring the rivet, still on the end of the bar, up to almost welding heat, then push it through the two holes and twist off.

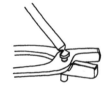

Drive the rivet into place over the pritchel hole and with a few heavy blows spread out the head.

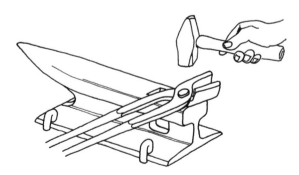

Quickly turn the two halves of the tongs over, taking care that the rivet does not fall out of place. With firm accurate blows spread out the head on the other side of the rivet.

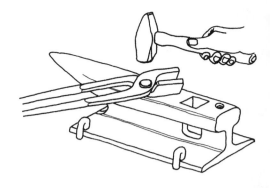

Forge the jaws of the tongs on to a piece of round section steel. A piece of straightened coil spring or the body of your round punch is ideal.

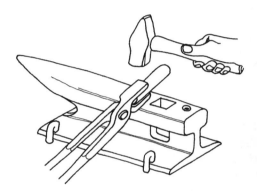

The inside of the jaws should now look like this.

With a small piece of bar holding the reins apart, forge the jaws down onto a piece of leaf spring. Try to use a piece the same thickness as the metal you work with most often.

Finally, quench the tongs in water, opening and closing the reins as they cool down. This will make sure that the tongs can move freely once cold. DO NOT quench tongs made with reinforcing bar as this would make them brittle.

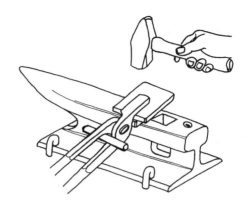

Simple fullers

Sometimes it is useful to be able to forge a groove in a piece of steel accurately. This can easily be done with a simple top and bottom fuller.

A simple bottom fuller can be made by bending a short length of mild steel into a **Z** shape, as shown.

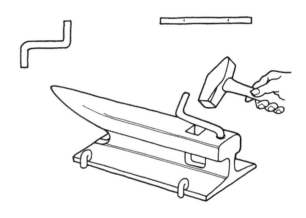

A straight piece of the same diameter steel can be used as a matching top fuller. The two tools are normally used together as shown.

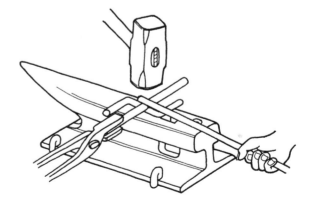

Hammer making

A range of very useful hammers can be made from old vehicle half-shafts. The most useful type for blacksmithing work is the cross-peen hammer. This section explains how it is made from start to finish. At the end of the section you will find drawings and text explaining how you can vary the techniques in order to make two other types of hammer.

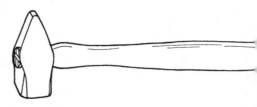

Cross-peen hammer

Before you can make hammers, two other tools have to be made: an eye chisel and an eye drift. The eye chisel is used to cut a slot in the hammer head. The eye drift is then used to open up the slot to the right size and shape for the handle to be fitted.

Eye chisel

The best material for making an eye chisel is medium- to high-carbon steel between 20 and 30 mm diameter. A straightened-out piece of heavy coil spring, an old anti-roll bar or a thin half-shaft would be ideal. When finished, the chisel should look like this.

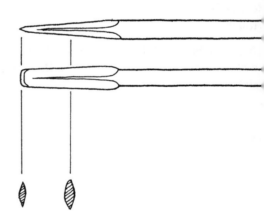

Form the striking area of the chisel by drawing down and flattening back the end of the bar.

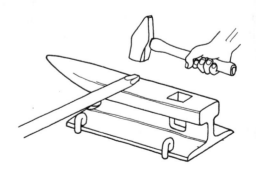

With a hot set or hot chisel, cut the shaft to length. The body of the chisel being forged should be fairly long (about 400 mm). This allows the hand holding it to be kept at a safe distance from the hot metal when it is used.

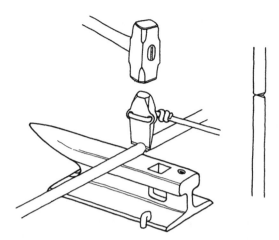

Draw down the blade section to a flat, even taper.

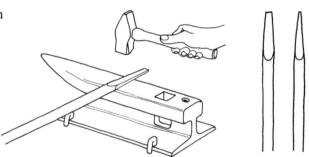

By working first along one side then the other, forge the whole length of the tapered area to a diamond shape as shown.

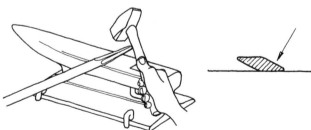

When looked at from the front and the side the chisel should now look like this.

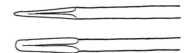

Grind or hot-file the blade as shown in the drawing.

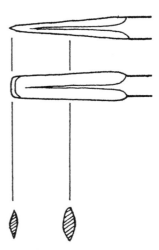

Eye drift

The eye drift should be made from a slightly thicker piece of steel than the eye chisel. A length of vehicle half-shaft is ideal. When finished the drift should look like this.

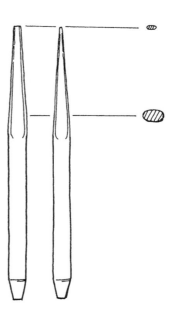

Form the striking area of the chisel by drawing
down and flattening back the end of the bar.

With a hot set or hot chisel cut the shaft to
length. The body of the drift should be a little
longer than the eye chisel. This allows the hand
holding it to be kept at a safe distance from the
hot metal even when it has been driven well
into the eye of the hammer.

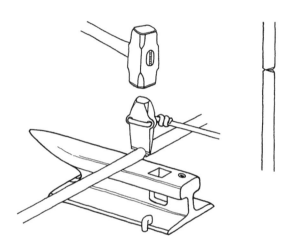

Draw down a flat, even taper of about one-third
of the length of the drift.

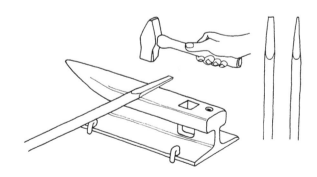

Forge in the corners by hammering in the
direction shown by the arrows in the drawing.
The finished drift should have a cross-section
the same shape as the eye (slot) in a hammer.

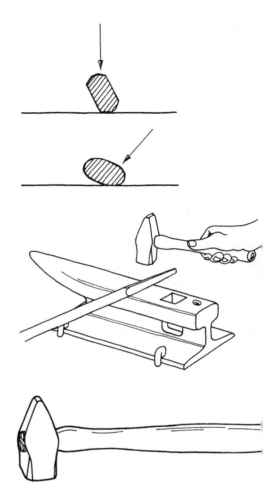

Cross-peen hammer

This cross-peen hammer is made from an old
vehicle half-shaft, i.e. medium-carbon steel.

Hammers can also be made from mild steel with
a piece of carbon-steel fire-welded to the face.
This process is called hard-facing and is described
at the end of this section.

Cut off the splined section of the shaft and draw
down a short, flat-sided taper.

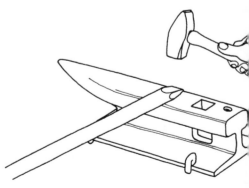

The end of the shaft will probably have become uneven and creased.

Cutting from both sides, remove the creased end.

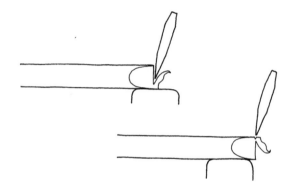

The workpiece should now look like this.

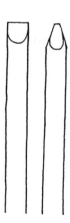

Finish off the peen by forging in the two long ridges left over from the last stage.

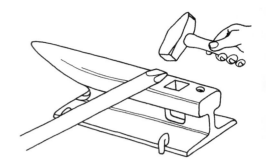

The finished peen should look like this.

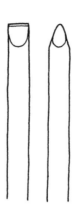

Having made the peen you can now start to form the eye.

Holding the shaft across the anvil with the peen horizontal, flatten a small area where the eye is to be punched. Before returning the metal to the fire, use the eye chisel to mark where the metal is to be cut. This mark must be exactly in line with the centre of the shaft.

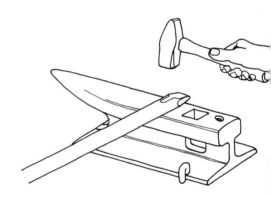

Re-heat the metal to its correct working temperature and drive the chisel a little over halfway through the shaft. Every few blows quench the tip of the chisel; this is extremely important as the heat from the shaft will very quickly soften the steel of the eye chisel. If you accidentally let the chisel get too hot be sure to let it cool to well below red heat before quenching or it will become brittle.

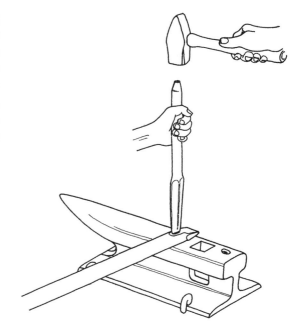

Each time the chisel is quenched it should be rotated half a turn before continuing the cut. This will make up for any tendency the chisel has to wander over to one side and will therefore help to keep the cut vertical.

Once you have cut a little over halfway through the shaft, turn it over and repeat the process from the other side.

Be very careful to start the second cut exactly in line with the first. This will help to ensure that the head of the finished hammer is correctly aligned when fitted onto its shaft. This time, drive the chisel all the way through until the slot is large enough to allow the drift to be used.

Working over the hardie hole (or over the edge of the anvil if there is no hardie hole) drive the drift through, first from one side then the other, until an eye is formed a little smaller than is needed to fit the hammer shaft. It is not necessary to quench the drift.

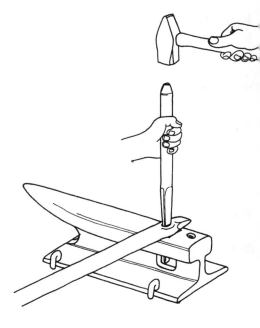

The workpiece should now look like this.

The metal which has been pushed out on either side of the eye should now be forged in to form the cheeks. When the hammer is finished these cheeks will help to ensure a good grip between the head and the hammer shaft.

During this stage no time must be wasted between driving in the drift and forging the cheeks. This is because the cheeks will cool down very quickly as heat is transferred to the drift. As soon as the metal is up to its correct working temperature remove it from the fire and drive in the eye drift. Lay the shaft across the anvil as shown in the drawing and forge out the side of the cheek nearest to the top of the drift. Quickly turn the metal over so that the drift is now pointing the other way and again forge out the side of the cheek nearest to the top of the drift. Remove the drift. The shaft should now look like this.

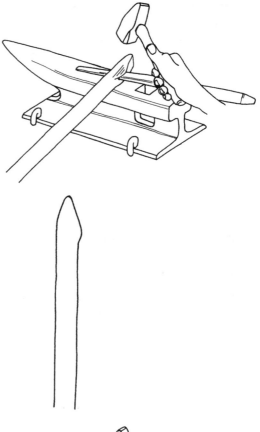

With the shaft once again brought up to its correct working temperature drive the drift in from the other side. This should be done over the edge of the anvil so that the newly forged cheeks are not damaged by the corners of the hardie hole.

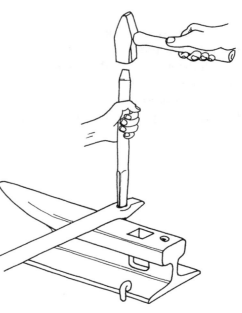

Forge out the remaining cheeks using the same method as before and remove the eye drift. The finished eye should now look like this.

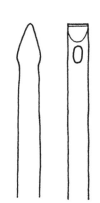

The next stage is to forge in the sides, top and bottom of the hammer head, along the rest of its length.

The hammer head is now ready to be cut away from the remaining half-shaft.

While there is still some heat left in the metal from the last process, use a hot set to mark a line all the way round the hammer head at the point to be cut. The distance between the eye and the face is normally the same as the distance between the eye and the end of the peen.

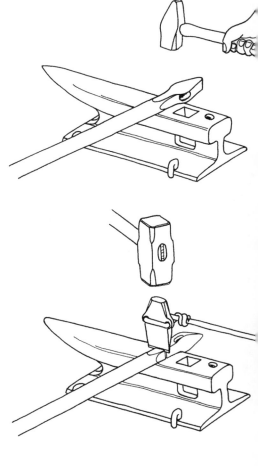

Cut the hammer head off from the shaft using a hot set and turning the shaft every few blows so that the cut finishes at the centre of the shaft.

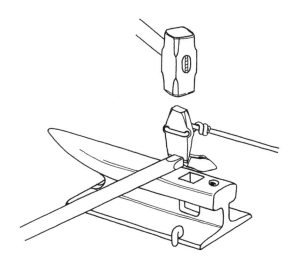

The hammer head should now look like this.

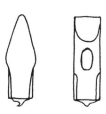

Holding the hammer head through the eye with a pair of tongs, remove the bulge in the middle of the face using a hot set.

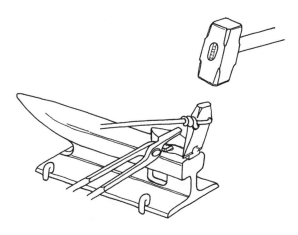

To do this you will need to hold the set at a slight angle.

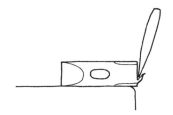

The face of the hammer can now be forged until it is flat and even. While you are forging the face, the peen should be supported on a block of wood to prevent it from being damaged.

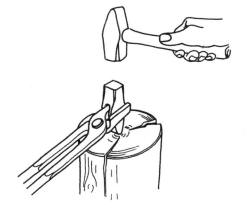

The hammer head is now ready to be annealed. This process is extremely important and if not done the head is likely to crack during the hardening process. To anneal the hammer head, slowly heat it up to a red heat and bury in dry ashes for several hours.

With a file clean the sides, the face and the peen of the hammer head and remove any sharp edges from around the face. The hammer head is now ready to harden.

Hardening and tempering hammers

There are several ways to harden and temper hammers. Three methods are described in this section. Do not attempt this process until you have read the section on heat treatment on page 21.

Method 1

For this method you will need to make tempering irons. These are lengths of mild which are forged to the same shape and si the eye drift and are used in the temp process.

Heat both the tempering irons up to a yellow heat. At the same time heat the hammer head up to a dull red heat. Remove the hammer head from the fire and, holding it with a pair of tongs through the eye, quench alternately the face and the peen in water until they are cool enough for water to stay on them without evaporating. During this process keep turning the head over, never spending more than a couple of seconds quenching the face or the peen. Try not to let the water get into the eye of the hammer as this section should not be hardened. The hammer is now ready for tempering.

Quickly clean the sides, peen and face of the hammer with a piece of old grinding wheel so that the temper colours will show up.

Remove one of the tempering irons from the fire and insert it through the eye of the hammer. The heat from the iron will now start to heat up the hammer head. As this happens temper colours will start to appear on the sides of the hammer. Each time the tempering iron cools to below red heat, it should be removed, placed back in the fire and replaced with a hot one. As the heat spreads, the temper colours will move towards the peen and the face.

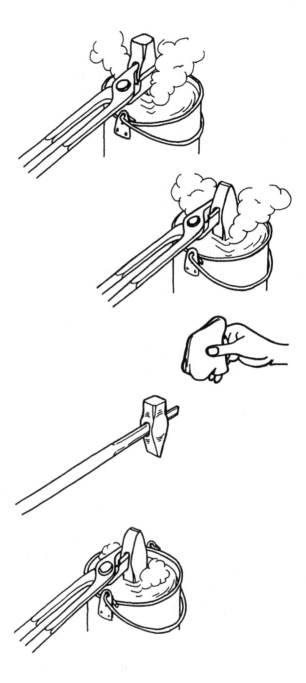

A well-tempered hammer should show a straw or dry grass colour on the face and a slightly darker brown on the peen. Keep checking the face and the peen, and as soon as the correct colour shows at either place quench that end of the hammer. Keep watching the remaining end of the hammer and as soon as the correct colour shows there, start turning the hammer, quenching first one end then the other until the whole head has completely cooled down. (Do not put the whole head into the water to cool it, only the face and the peen.)

Method 2

For this method you need a can with a small hole punched in the bottom and two buckets of water. The can should hold about half to three-quarters of a litre.

Clean the hammer as described in Method 1 and bring slowly up to a dull red heat. Dip the peen of the hammer in one of the buckets of water, and at the same time quickly fill the can with water from the other bucket. Let the water drain out of the hole on to the face of the hammer, keeping your hand clear of the steam. As soon as the can is empty, remove the peen from the water and quickly clean the sides, peen and face so that the temper colours will show up. The eye section of the hammer will still be hot and the temper colours will start to move towards the peen and face. The face should be quenched when dry grass or straw colour is reached and the peen should be quenched at a slightly darker brown.

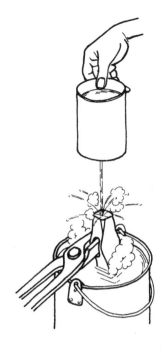

Method 3

Certain steels have a tendency to crack during the hardening process, even when the item made has been carefully annealed. With these steels oil-quenching, which is a slower process, can sometimes help. To do this you need a large container with about five litres of used engine oil in it, a piece of steel wire to hang the hammer head on, and a piece of scrap sheet steel large enough to cover the top of the container. The cover will prevent the oil from burning during the hardening. The process should be done in a well-ventilated area.

Bring the hammer head up to a dull red heat. While it is heating bend the end of the wire into a hook shape. Carefully lift the hammer h from the fire on the end of the wire hook lower it into the oil. Immediately cover container with the sheet of steel and leave hammer head to cool down. When cool, hammer head is hardened. Tempering is c in exactly the same way as described in Met 1. Always allow the oil to cool down be using it to harden further items, and alway sure that you have enough oil in the contai

A good substitute for oil in rural areas ca made by mixing soil or clay with water. thicker the mixture, the slower the coc process will be.

The hammer shaft

A well-made hammer is not complete until it is properly fitted to a well-made shaft. The type of wood which is used in your area for making axe handles will probably be ideal for the hammer shaft. Choose a piece with the grain running as straight as possible along its length. The first stage is to cut the timber to length.

Using an adze reduce the timber to a rectangular section as shown.

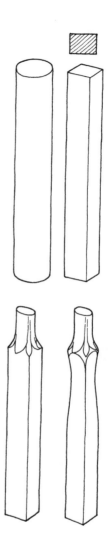

Carefully trim down the end of the timber with the adze until it fits tightly into the eye in the hammer head.

Thin down the handle for a short length just behind where the head will be fitted to the shaft. When the finished hammer is in use this thinner part of the shaft will help to soak up the shock waves from each hammer blow, and will make the hammer much less tiring to work with.

The final shaping can now be done using an adze or a rasp. Two views of the finished shaft and cross-sections of both ends are shown.

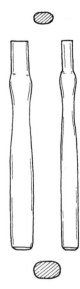

A slot should now be cut in the end of the shaft as shown. It should be approximately three-quarters of the depth of the eye.

The next stage is to make a wooden wedge. Choose a piece of wood with the grain running as straight as possible along its length and cut it to shape as shown. The width of the wedge should be the same as the width of the slot in the end of the hammer shaft.

The shaft can now be fitted into the eye of the hammer and the wedge driven in as shown.

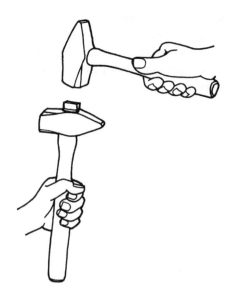

As the wedge is driven in, the end of the shaft will expand to fill the eye of the hammer completely. The wedge can now be cut off as shown, and finally driven in until level with the end of the shaft.

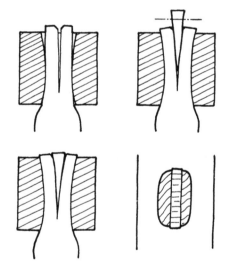

Finally, a metal wedge should be made and fitted to secure the head firmly on to the shaft. The drawings show the stages. The barbs which stop the wedge from working loose are cut using a hot chisel.

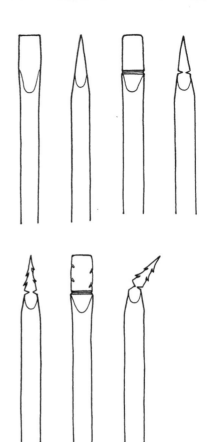

Drive the metal wedge in at right angles to the wooden wedge as shown.

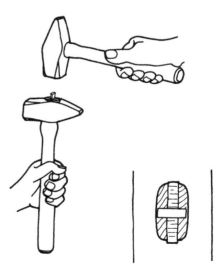

The ball-peen hammer

A ball-peen hammer is made in almost exactly the same way as the cross-peen hammer. The only difference is in the first few stages. The finished hammer should look like this.

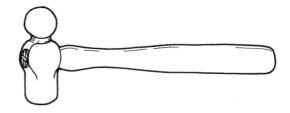

The first step is to forge a chamfer on the end of the bar.

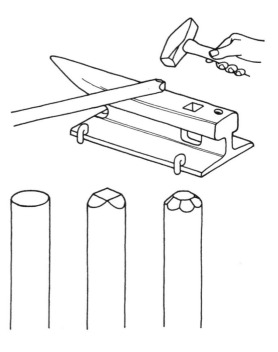

Using the simple fullers described on page 47 forge a groove as shown in the drawings.

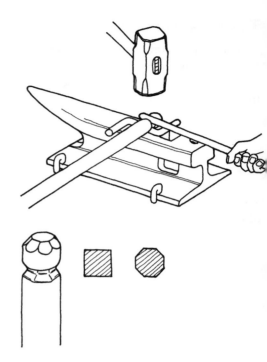

With the top of the ball resting on the anvil, round off the back of the ball.

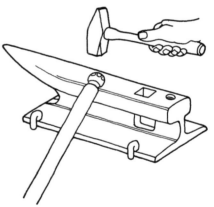

Working with the groove against the edge of the anvil, round off the top of the ball.

The finished peen should now look like this.

The rest of the hammer can now be made in the same way as the cross-peen hammer (see page 54).

The claw hammer

A carpenter's claw hammer can be made out of a thin piece of vehicle half-shaft. The finished hammer should look like this.

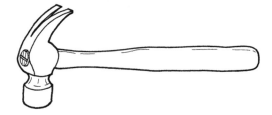

Hammering directly over the edge of the anvil, forge a step in the shaft as shown in the drawings.

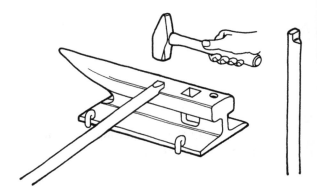

Continue to draw down the step until it is long enough to form the claw of the hammer.

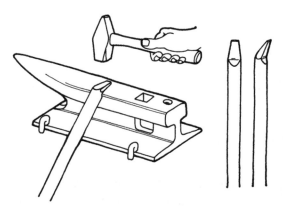

Using a hot chisel carefully mark out the centre line along the length of the claw.

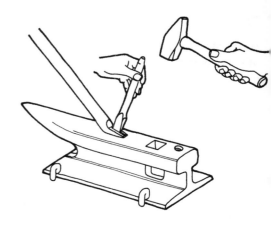

Cut most of the way through the claw using a steep-sided hot chisel.

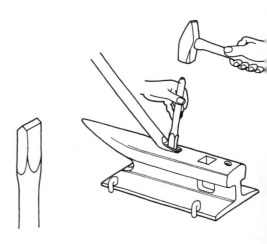

Using a piece of mild steel sheet to protect the face of the anvil, finish splitting the claw using a normal hot chisel.

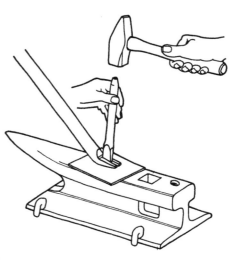

Gently open out the claw of the hammer using a hot set.

The claw should now look like this.

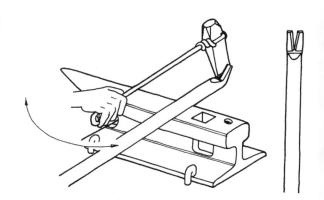

From this stage onwards the process is almost the same as has been described under the section on making a cross-peen hammer. There are, however, three differences.

o The claws should be forged into a slight curve just before the hammer is cut away from the rest of the half-shaft.
o A groove should be made between the eye of the hammer and the face using a simple top and bottom fuller.
o Finally, the claws should be tempered until pale blue. This should prevent them from breaking off during use.

Hard-facing a hammer

By using fire-welding techniques, hammers can be made from mild steel and given a hard, high-carbon-steel face.

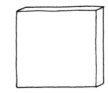

Using mild steel, follow the instructions for making a hammer as laid down in the previous pages until you reach the section on hardening and tempering.

Using a hot set cut a piece of leaf spring to the same size and shape as the face of the hammer. This is best done in two stages: first, cutting out a square, and then removing the corners. The finished piece should look like this.

A special nail now has to be made to hold the piece of leaf spring in position during the welding process.

Straighten out a short piece of coil spring and draw down a square point.

File or forge the tip so that a point is formed at one side.

The nail is finished by cutting off the drawn-down section as shown.

Using a round punch make a hole in the centre of the piece of leaf spring. The hole should be just large enough for the nail to be driven into.

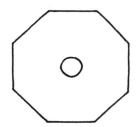

Heat up the hammer head and, using a small punch, make a hole in the centre of the face. As the punch is very small it will need to be quenched after each blow of the hammer. This stage should be done with an assistant holding the hammer head over the wooden block as shown.

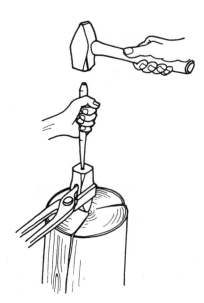

In cross-section the hammer head should now look like this.

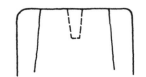

Bring the hammer head up to a bright yellow heat, then drive the nail through the leaf spring and into the hammer head with rapid firm blows. Both the nail and the leaf spring should be cold at this stage.

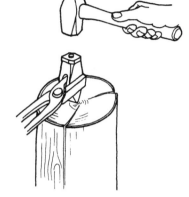

As the nail is driven in it will hook over and firmly pin the face on to the hammer.

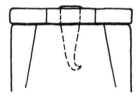

While the hammer head is still hot, return it to the fire and bring up to welding heat. Sprinkle a little clean river sand between the two pieces of steel. The sand will melt and act as a flux as the two pieces heat up. Bring the hammer head back up to welding heat and weld on the face using firm rapid blows, starting from the centre and working outwards until the whole face is welded on. Hot flux and slag will be driven out as the weld is performed, so make sure that nobody is in danger of being burned by it.

Heat the hammer up to red heat and anneal in ashes.

With a file clean the sides and face of the hammer head and remove any sharp edges from around the face. The hammer head is now ready to harden.

The hammer face can be hardened and temp using the techniques described on pages 6(The peen, being of mild steel, canno hardened and therefore need not be quen(

5. PRODUCTS

This section contains instructions on how to make a range of products. The designs chosen are based upon those seen by the authors in Zimbabwe and Malawi. Designs of products vary from country to country and even between regions within a country. The following instructions should therefore not be used as a fixed set of rules, but rather as a guide to techniques.

Axe-making

Most blacksmiths find that there is a demand for axes. The following pages will show you how to make an axe head from start to finish. They will also show how you can save fuel and time by making several items at once.

Axes should be made from high-carbon steel. Used vehicle leaf springs are ideal. The width and the thickness of the metal chosen, and the lengths of the pieces cut off, will depend upon the weight and size of axes wanted.

Heat up a length of leaf spring and, using a hot set or chisel, cut a little over halfway through the metal as shown.

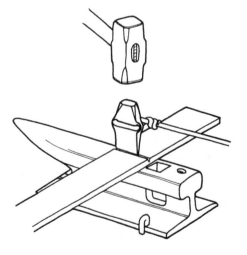

Turn the metal over and with the cut line level with the edge of the anvil bend the metal downward slightly. When broken off this piece will provide the metal for two axes.

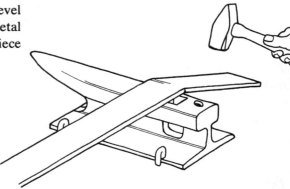

Heat the metal up further along so that a second piece the same size can be cut. Using the hot set, make the second cut.

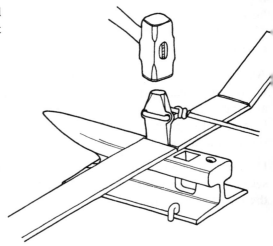

Where the first cut was made the metal should now have cooled enough to be broken off. To do this, turn the metal over, lay it along the anvil and give it a few sharp blows with the hand hammer.

The metal at the second cut line can now be bent slightly and the process repeated until the whole leaf has been cut up.

When making a large quantity of axes it will save fuel and time if metal is stacked up in the fire. A small amount of fuel should be put between the layers of metal.

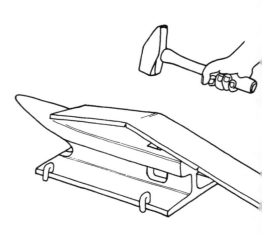

Always take the bottom piece of metal from the fire as this will be the hottest. When the bottom piece has reached forging temperature remove it from the fire and using a hot set cut a little over halfway through as shown. The cut should start at one corner and go about halfway towards the opposite corner.

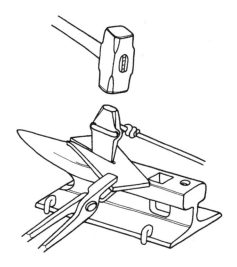

Turn the metal on its side and strike the edge a few times with the sledgehammer. This will bend the metal along the cut line.

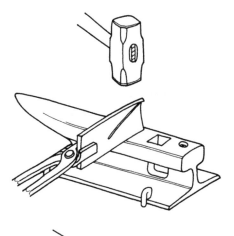

Turn the metal round in the tongs and continue the cut from the centre to the other corner.

Once again turn the metal on its edge and give it a few blows with the sledgehammer. The metal should now have a bend all the way along the cut line. Put the metal aside to cool down and start working on the next piece. Repeat the process until all the pieces of leaf spring have been cut from corner to corner.

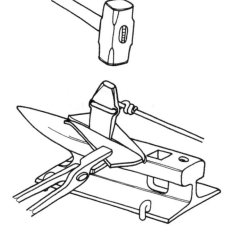

Starting with the coolest of the pieces, use a
hand hammer to break each piece in half along
the cut line as shown.

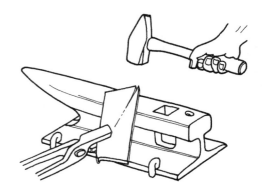

Each piece of metal should now look like this.

The pieces of metal should now be stacked up
in the fire as shown in the drawing.

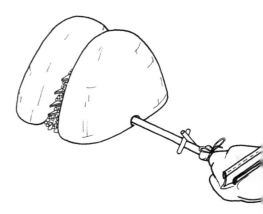

When the bottom piece has reached the correct
temperature, remove it from the fire.

The blade section now needs to be cut to shape. To do this, cut the metal most of the way through as shown using a hot set. A bigger piece should be cut from the longer side.

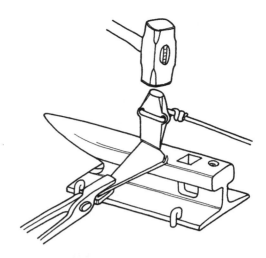

The axe head should now look like this.

Holding the metal in a pair of tongs, carefully quench the corners of the axe head as shown and break them off along the cut line with a few blows of the hand hammer.

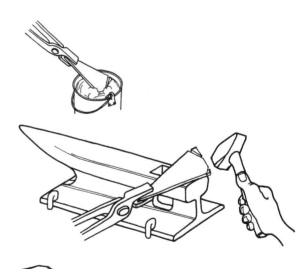

The first axe head should now look like this.

All the axe heads should now be brought to this stage, returning each one in turn to the top of the pile ready to heat up for the next stage.

Working near the edge of the anvil start to draw down the cutting edge of the blade.

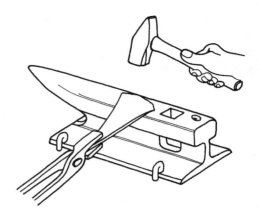

As the blade is drawn down it will start to spread out sideways. If you wish to produce an axe with a straight edge you should turn the axe on its side and forge the sides back in.

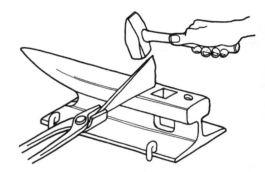

Before the blade becomes too thin it should be rounded off with gentle blows of the hand hammer.

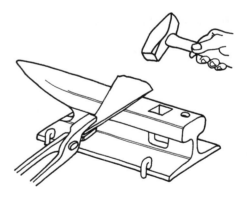

The drawing down of the cutting edge can now be finished off using rapid, slightly angled hammer blows along its length.

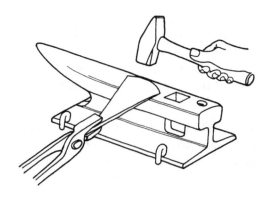

At the end of this stage each piece should be turned round and returned to the top of the pile of axes in the fire so that the other end of the axe head will heat up. Each piece of metal should be worked on in turn, always starting with the bottom piece. The tang is drawn down by forging in the sides of the axe until the correct width is reached and then drawing down the tip to a square point.

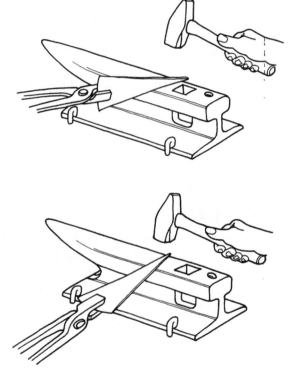

Each of the finished axe heads should look like one of the two types shown. The drawing on the left shows an axe with the sides forged flat. The right-hand drawing shows one where the sides of the blade have been allowed to spread out.

Axes can be hardened and tempered to help them stay sharp for longer. However, if hardened, they cannot be sharpened using a file as is common practice in some areas.

Hoe-making

Two methods of making traditional pattern hoes are described in the following pages. The first is made in two parts, the blade being cut from sheet steel and the tang forged from a piece of round bar. The second type is forged from a single piece of steel cut from a used tractor plough disc. If you have access to tractor plough discs the second type is simpler to make and is usually slightly stronger. This is because the plough discs are normally made from medium-carbon steel.

Two-piece hoe

To make the hoe you will need three different types of raw material. The tang is made from round section mild steel or reinforcing bar (approximately 20 mm diameter); the blade is cut from sheet mild steel (approximately 3 mm thick); and the rivets are made from a small-section, round mild steel bar (6–8 mm diameter).

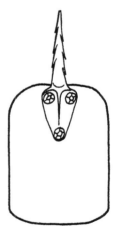

The finished hoe should look like this.

To forge the tang take the large-section mild steel bar and draw down the end to a square point.

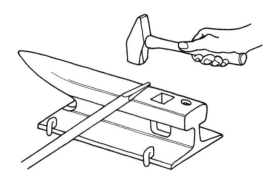

Forge the point slightly off-centre so that one of the flats is in line with one side of the bar.

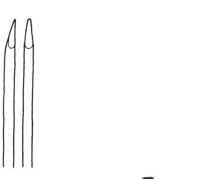

Hold the metal across the anvil with the flat side downwards and start to spread the drawn-down section sideways leaving a ridge down the centre line.

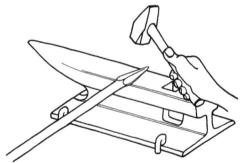

The workpiece should look like this.

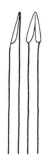

Turn the metal over and continue to spread the metal out sideways. Keep the ridge just off the anvil so that it is not spoilt as the sides are spread out.

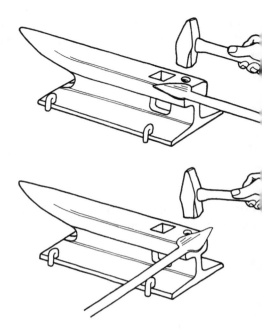

The piece of metal should now look like this.

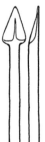

Using a hot set or a hot chisel cut the workpiece off the bar leaving enough metal to draw down the tang.

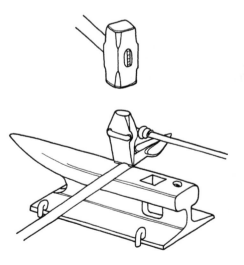

Draw down the tang to a square point.

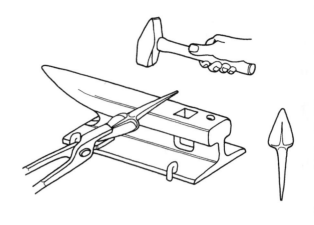

Punch three holes for the rivets. The holes should be the same diameter as the metal from which the rivets are to be made.

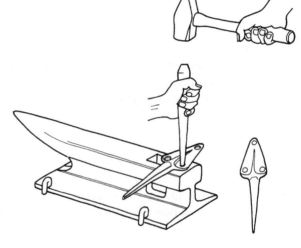

Turn the metal round and use a chisel to cut barbs along the tang. These barbs will help to hold the hoe firmly on to its handle.

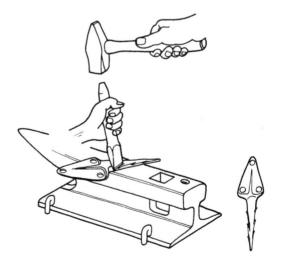

Using a cold chisel, cut out a rectangle of sheet steel for the hoe blade.

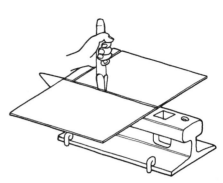

Again using the cold chisel, cut the blade to shape. Remember that the picture is only a guide and the shape will depend upon the type of hoe preferred in your region.

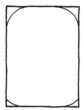

Take the tang you made earlier and use it to mark the position for the holes in the hoe blade.

Punch the holes ready for the rivets. This can
be done without heating the metal.

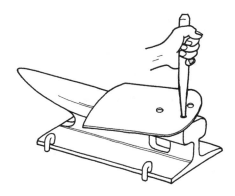

Chamfer the end of the smaller diameter mild
steel bar ready to make the rivets.

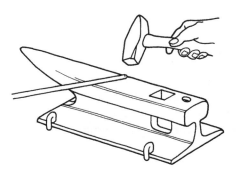

Mark off the metal, allowing about one and a
half times the diameter of the rivet on each side
of the two pieces to be joined.

Using a cold chisel, cut in towards the centre of the bar until the metal is almost cut through, and then break the piece off using a hand hammer over the edge of the anvil.

Using the same techniques, prepare two more rivets.

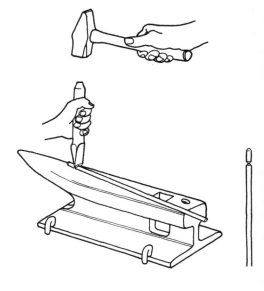

Small rivets of this type can be made and fitted cold. Line up the holes in the tang with the holes in the blade and drive the first rivet into position over the pritchel hole in the anvil.

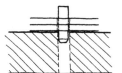

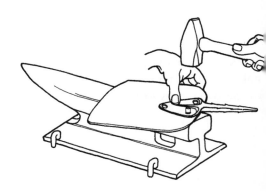

Rivet the two pieces of metal together by first hammering vertically on to the rivet and then by working at an angle around the edges.

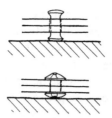

Finally the blade can be bent to shape.

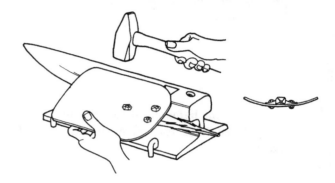

One-piece hoe

This pattern of hoe is easier to make, if you have access to used tractor plough discs. You will need a fairly large fire for the first stages of the job. Up to eight hoes can be made from a single disc.

The finished hoe should look like this.

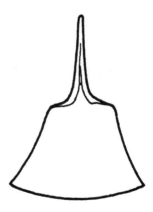

Heat up an area of the disc from the centre to one edge. Using a hot set, cut a straight line almost through the metal between the centre and the edge. Lay a piece of scrap mild steel under the plough disc to protect the anvil and the set blade during the cutting process. Return the metal to the fire and heat it up so that a second cut can be made as shown. Repeat the process until the whole disc has been divided into sections.

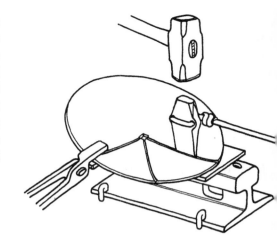

Using a hand hammer, break the pieces off along the cut lines. Each piece should now look like this.

Draw down the tang to a gently tapering square point.

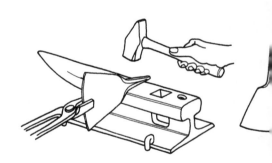

Finish off the hoe by forging the blade section.

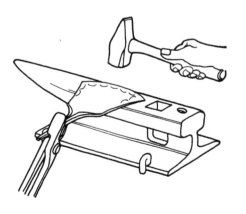

Knife-making

Many different shapes and sizes of knives can be made using blacksmithing techniques. This section describes, step by step, how to make one of them. With experience you will be able to use the same techniques to make a wide range of knives.

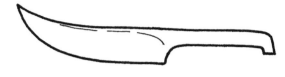

To make a tough knife which will also stay sharp, medium- to high-carbon steel should be used. For heavy knives, thin pieces of vehicle leaf spring are ideal. Very good small knives can be made from old saw blades. The knife described is made from a piece of vehicle leaf spring. When finished it should look like the drawing here.

Working with the metal held on its edge across the anvil, draw down enough metal to form the handle. Allow a little extra length so that the tip of the handle can be bent over (see drawing above).

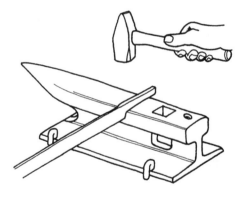

Forge in the sides of the handle until they are flat and slightly thicker than the blade section.

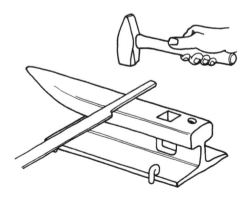

Using a hot set, cut the leaf spring across at an angle as shown.

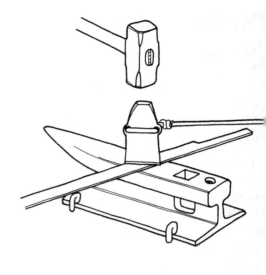

The workpiece should now look like this.

With the blade facing downwards, work along the back of the knife until a smooth curve is formed. Occasionally lie the blade on its side and flatten it.

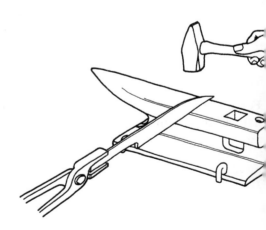

With the knife held blade downwards across the anvil, forge a bend along the length of the blade. During this process make sure that the knife if firmly held in the tongs.

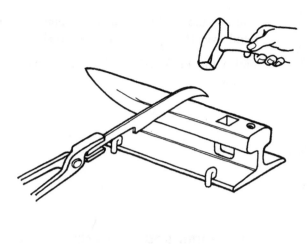

The knife should now look like this.

Hammering at a slight angle, work along the length of the blade as shown. As the blade is forged the knife will start to straighten out.

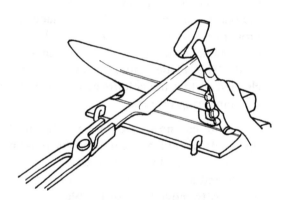

Continue forging the cutting edge of the blade until it is thin enough to finish off with a few strokes of a file.

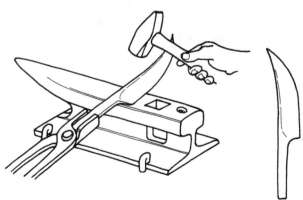

Holding the blade carefully in a pair of tongs, bend over the tip of the handle as shown.

Finish the knife off by sharpening the blade with a file.

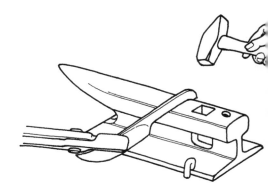

Hardening and tempering knives

Knife blades will remain sharp for longer if they are hardened and tempered. The easiest way to do this is use the two-heat process. Do not attempt this process until you have read the section on heat treatment on page 21. First the blade is annealed. The entire blade is then brought up to a dull red heat and the cutting edge quenched until cold. This will make it very hard. The blade should now be carefully cleaned so that the colours can be seen during the tempering process.

The knife can now be tempered. The back of the blade should be heated up by pressing it against a piece of red-hot metal. As the heat is conducted across the blade the temper colours will start to move towards the blade edge. The blade should be left in contact with the metal until a dark purple colour is seen all along its edge. As soon as this happens it should be quenched until cold.

The sickle

A useful sickle can be made from 10–12 mm round section mild steel bar. If this is not available, flat bar 5 mm by 20 mm can be used. When finished the sickle should look like the drawing here.

Draw down a gentle square taper on the end of the bar to form the tang. Offset the tang slightly by working with the hammer directly over the edge of the anvil as shown. The tang should be made fairly long as it will have to pass right through the handle and still be long enough to be bent over at the end.

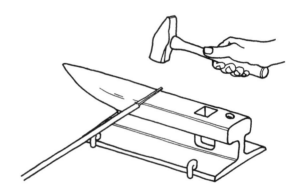

The finished tang should now look like this.

With the flat side of the tang facing downwards, bend the tang over the edge of the anvil.

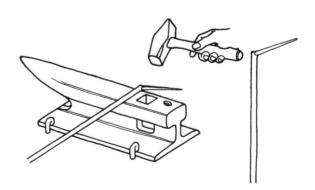

The next stage is to start forming the curve of the sickle. As the blade is forged the sickle will tend to straighten out, so the curve should at this stage be made much tighter than is wanted on the finished item. Start by bending the blade section at the point nearest to the tang.

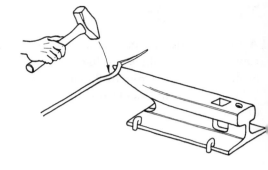

Flatten out the whole of this section including the tang.

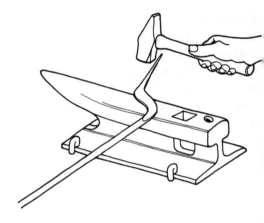

The metal can now be cut off using a hot set as shown.

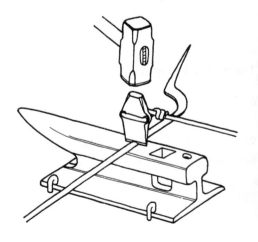

Draw down the tip of the blade section to a square point.

The sickle should now look like this.

Working over the bick of the anvil, bend the blade section round until it almost forms a complete circle.

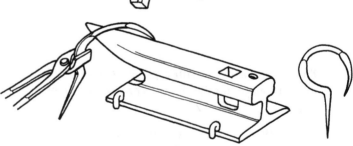

Using a hand-hammer, flatten the blade.

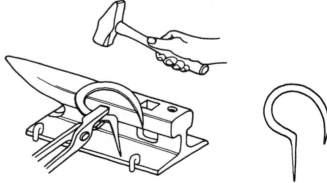

Using slightly angled hammer blows, forge down the edge of the blade until it is thin enough to be sharpened with a few strokes of a file. As you do this the sickle should open out to a nice gentle curve.

The sickle is now ready to have the teeth cut into the blade.

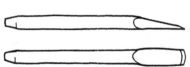

There are several ways of cutting the teeth. One system is to forge a one-sided cold chisel as shown and to use this to cut the teeth.

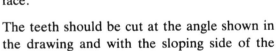

The cutting is done with a piece of scrap mild steel placed under the sickle to protect the anvil face.

The teeth should be cut at the angle shown in the drawing and with the sloping side of the chisel facing towards the tip of the sickle.

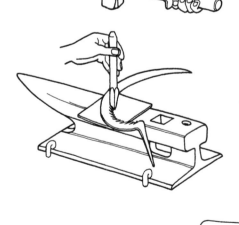

The teeth can also be cut using a hacksaw or a three-square (triangular) file.

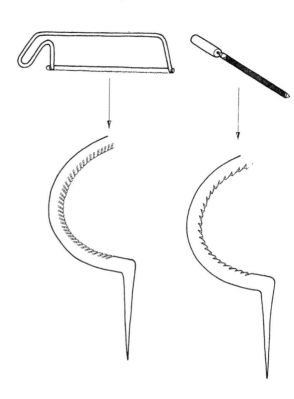

The sickle is now ready for the handle to be fitted. Select a suitable piece of wood, and cut it square on one end and at an angle on the other, as shown.

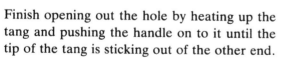

Draw down a point on a thin piece of mild steel bar. Use this to burn a hole through the handle for the tang.

Finish opening out the hole by heating up the tang and pushing the handle on to it until the tip of the tang is sticking out of the other end.

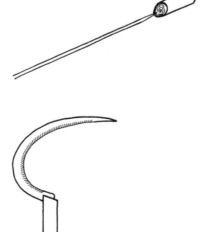

Remove the sickle from its handle and heat up the tip of the tang. Push the handle firmly on to the sickle and quickly bend the tip of the tang over as shown.

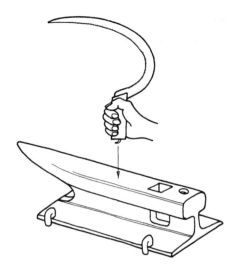

Finally, quench the handle to prevent it from being burnt.

6. SETTING UP A WORKSHOP

For a blacksmith to work effectively, some kind of workshop is needed. This need not be an expensive building. A small corrugated metal roof raised up on poles is good enough, as long as the following rules are obeyed:

o The roof must be high enough to give plenty of space for working under; remember that when you are striking metal hard the hammer is often raised well above the head.
o The roof must be arranged so that direct sunlight does not fall on the hearth or the anvil as this would make it difficult to judge the temperature of the metal while working.
o Any poles used should be placed outside the working area so as not to get in the way.

If an enclosed workshop is to be used, good ventilation is necessary and a chimney should be provided over the fire to take away smoke and fumes.

Three pieces of equipment are needed to set up a blacksmith's workshop. They are the anvil, the bellows, and the hearth.

Anvil

Factory-made anvils are expensive and often impossible to find. Many alternatives are available in the form of broken or worn-out parts from machines. When you are trying to decide what to use as an anvil the following points may be helpful:

o The item should be tough, so avoid things made from cast iron which may break in use.
o Pick something heavy; a light anvil will be tiring to work on and will bounce around, no matter how firmly it is nailed down.
o Try to find something which has a flat area on top, with one rounded and one square edge. It will also help if there is a hole which can be used for punching over.

An excellent anvil can be made from a length of old railway line if you have access to a workshop with cutting equipment. The piece of railway line should be approximately 500 mm long, and should be cut along the lines shown in the drawing below.

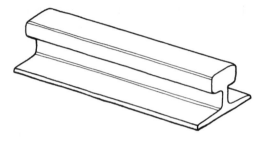

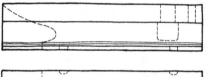

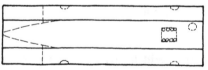

Broken lines indicate where cuts should be made

A 35 mm square hole should be cut from the top through to the hole in the web, to form the hardie hole. The cutting can be made easier if holes are drilled as shown in the drawing before the gas cutter is used. Slightly behind this and to one side, a 10–12 mm hole should be drilled, to form the pritchel or punching hole.

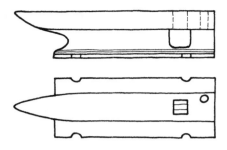

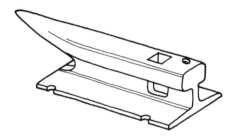

Finally, the anvil can be finished by grinding or forging the bick to shape as shown.

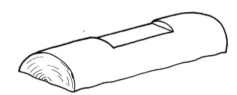

Two tried and tested systems for mounting the anvil are shown below. The first is ideal for those who prefer to work sitting on the ground. A split log is used with a flat area cut out to fit the base of the anvil.

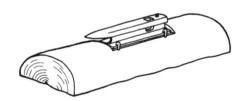

The anvil is held firmly in position with metal spikes.

For those who prefer to work standing up, the anvil should be mounted so that it is set at a comfortable height to work on. A good heavy piece of tree trunk makes an ideal support.

The mounting spikes can easily be made from short pieces of scrap mild steel or reinforcing bar. The five stages of making the spikes are illustrated here.

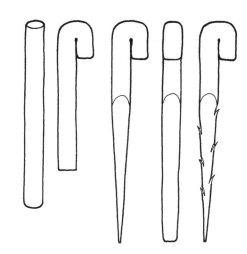

Bellows

There are many different types of bellows which can be used by a blacksmith. Four different types of bellows are covered in this section with a detailed description of how to construct and use one of them, the goat skin. When compared with most other types of bellows, the goat skin will be found to be easier to make, more reliable and less expensive. For these reasons most attention has been given to it in this book.

Goat-skin bellows

Goat-skin bellows have been in use for over 2000 years in Africa, for both smelting and forging iron. The simplest type of goat-skin bellows is the single-skin bag. To make this type of bellows a large goat should be chosen. In some districts female goats are preferred as it is thought that their skin is softer after it has been treated.

After slaughtering the goat, hang it from a tree and cut the skin as shown by the dotted lines in the illustration.

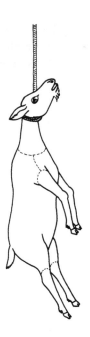

Remove the skin, being very careful not to tear or cut it in the process.

The skin, which is now inside out, should be hung up in the sun as shown. Tie a piece of string round the two openings where the back legs were and fill the skin with earth or sand.

The skin can now be scraped clean, oiled and worked until supple. The scraping should be spread over several days, starting as soon as the skin is hung up and not finishing until the skin has dried out. At this stage the skin should be oiled; groundnut oil is ideal. After oiling, the skin can be beaten with sticks which will help to soften the leather. Further oiling and beating should take place until the leather is soft and flexible.

The next stage is to fit a pipe on to one of the leg holes. A piece of old water pipe or some tubing from a broken cycle frame is ideal. The other leg should be left tied up. Two flat strips of wood with holes in can now be sewn across the opening in the bag, as shown in the drawing.

When using the bellows, the top bars must be held together on the down stroke, forcing the air out along the tube and into the fire. The top bars should then be separated and raised allowing the bag to fill with fresh air and the whole process repeated. It is important to make sure that the top bars are separated before the upstroke begins or fire will be drawn back into the bellows and they will soon become spoilt.

One alternative type of handle can be made by sewing round sticks into the top of the bag and cutting slots in the outer layer of leather for the fingers and thumbs of the operator.

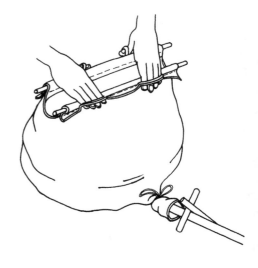

By making only one set of slots you can make bellows to be operated single-handed. Bellows of this type are often used in pairs with two pipes leading into the hearth. This arrangement, if used correctly, gives a constant and even draught to the fire.

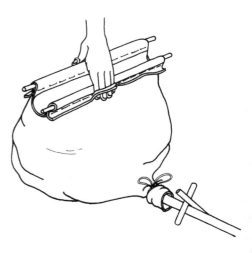

Rotary blower

The simple rotary blower works well on both coke- and charcoal-fired hearths. It is more costly than the goat skin and requires some tinsmithing skills to make it. If made well it is easier to use than the goat skin and will bring a fire up to full heat more quickly than the water bellows. It requires some maintenance, however.

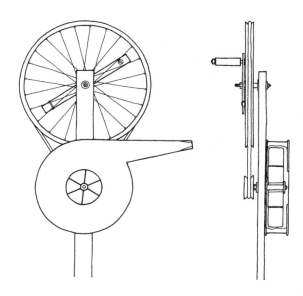

As can be seen from the cut-away drawing, the blower is worked by turning a handle which is fixed on to a bicycle wheel. The blower itself contains a six-bladed fan made from sheet steel. The fan is mounted on a bicycle wheel spindle. On the other end of the spindle is a small pulley wheel which is linked to the driving wheel by a rubber belt which could be made from an inner tube. As the fan turns, air is drawn in through the central hole and driven into the fire through the funnel-shaped pipe at the top of the blower.

Water bellows

The water bellows is a fairly recent design. It can only be made in areas where old oil drums are available and affordable. It works well on coke and charcoal fires and has the advantage that once it is built, little or no maintenance is required, especially if the drums are protected from rust with a good coat of paint.

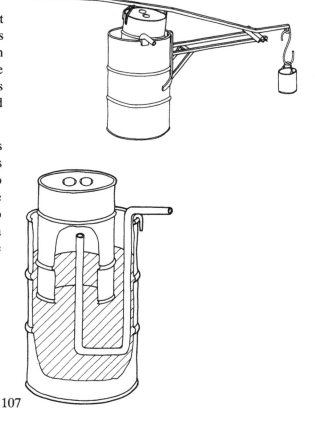

The cut-away drawing shows the working parts of the bellows. As the handle is lifted, air is drawn into the small drum through the two valves in the topl As the handle is lowered, the air is forced through the U-shaped pipe and into the fire. The water in the large drum acts as a seal, preventing the air from leaking out as the bellows are used.

107

Chinese box bellows

The Chinese box bellows have the advantage that air is driven into the fire as the handle is being both pushed and pulled. This makes them very efficient to use. Unfortunately this type of bellows is both expensive and difficult to make. The other disadvantage is that in some climates the timber panels may warp which could stop the bellows from working altogether.

The cut-away drawing shows the inside of the box bellows. The handles are attached to two lengths of wood, which can slide freely through the ends of the box. Inside the box they are fixed on to a wooden panel which is moved from end to end of the box as the handles are moved. As the panel is moved to the right, air is drawn into the left-hand section of the box through a simple valve in the end of the box. At the same time, air is being pushed from the right-hand side of the box through a tapered channel in the front section of the box and out through a simple valve in the front panel. From here the air goes down a tube and into the fire. Moving the handles to the left will reverse the whole process, drawing air into the right half of the box and driving it into the fire from the left.

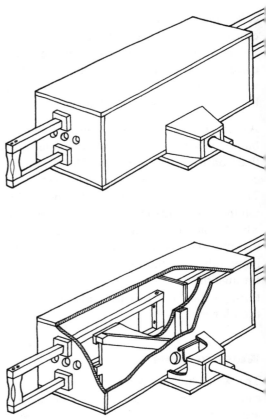

The hearth

In countries where most of the work is done using old leaf springs, a long narrow fireplace is often used.

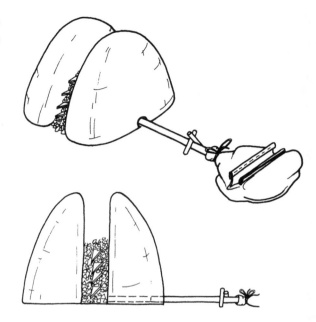

The fire is set between two mounds of termite clay with the air driven in through a hole at the base of one of them. This design reflects the heat back on to the work and is very efficient. Work can also be stacked in layers between the two mounds.

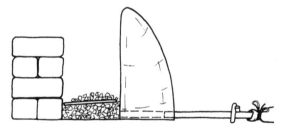

A fireplace which can be altered to suit the type of work in progress can be made by building a single fixed mound for the air to be driven in through. The second mound is replaced with a pile of mud blocks which can be moved to adjust the width of the fire.

Many blacksmiths find that they can work better standing up. Both the hearth and the anvil should therefore be raised off the ground to a comfortable working height. One method of raising the hearth is to make a solid box of rammed earth or mud blocks and to construct the hearth on top of this.

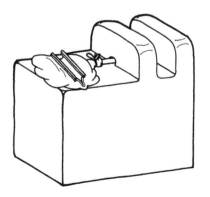

APPENDIX:
FORGED TOOLS FOR THE CARPENTER

These designs are based on the tools described in Aaron Moore's book, *How to Make Carpentry Tools*, published by Intermediate Technology Publications. These step-by-step instructions for making chisels, plane irons and a carpenter's brace and bit complement the requirements of that manual, but can of course be adapted as necessary.

Most of the items in this section are used as parts for wooden-bodied tools, and therefore have to be made accurately to the dimensions shown.

25 mm chisel

The 25 mm chisel is a useful tool for general-purpose carpentry work. It can also be used as a blade for the wooden rebate plane. The chisel is used in the same way as a factory-made chisel except that the handle is struck with a hammer rather than a mallet. When finished the chisel should look like this.

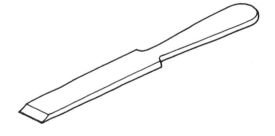

The drawings show the dimensions in millimetres of the finished tool.

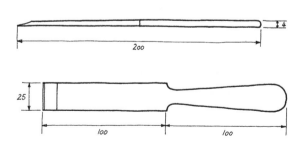

Using a hot set, cut out a rectangle of 4 mm leaf spring, 180 mm by 25 mm.

Make a chalk line halfway along the piece of steel and use this as a guideline when starting to draw down the handle.

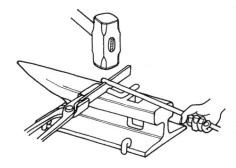

Using the simple top and bottom fuller described on page 47 make a groove on each side of the metal, level with the chalk line.

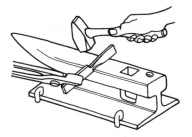

Working with the metal held at an angle as shown, draw down the handle.

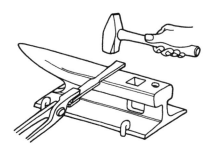

Flatten the handle of the chisel back to an even thickness.

The chisel should now look like this.

Holding the chisel firmly in a pair of tongs, round off the end of the handle.

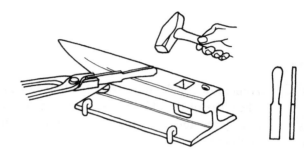

Draw down the blade to a gentle taper.

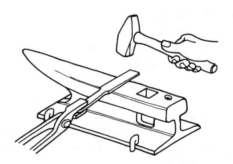

Forge the sides back in until they are parallel.

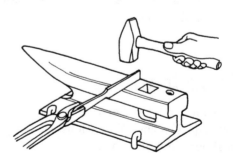

Working at the edge of the anvil with slightly angled hammer blows, draw down the cutting edge until the bevel is 10 mm long.

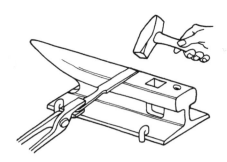

Anneal the blade and carefully file into shape. The sides of the chisel must be filed until they are square and straight and a smooth flat finish is obtained.

The tool is now ready to be hardened. Bring the whole of the blade to a dull red heat and quench half its length in water until cold. Polish the blade with a piece of old grindstone and temper to a pale straw colour. The easiest way to do this is to hold the chisel against a large piece of heated scrap steel. Keep the edge of the blade clear of the heat source. The temper colours will be clearly seen moving along the chisel as the heat is transferred. Re-quench the chisel as soon as the pale straw colour reaches the tip of the blade.

The blade can now be ground to the correct angle and sharpened.

Plane iron

The plane iron has been designed to fit the wooden jack plane. When finished the plane iron should look like this.

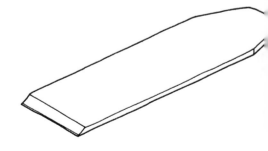

The two drawings here show the dimensions of the finished tool.

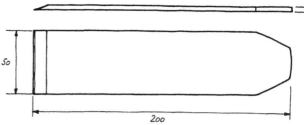

Using a hot set, cut a rectangle 200 mm by 50 mm out of a piece of 3–4 mm thick leaf spring.

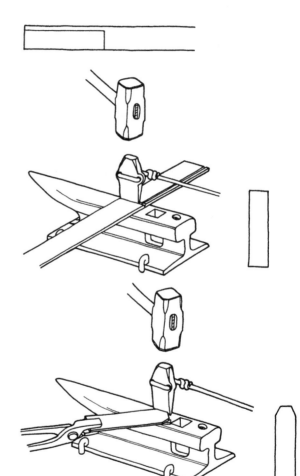

Holding the metal with a pair of tongs, cut off the corners as shown and square up the sides of the plane iron.

Working at the edge of the anvil and using slightly angled hammer blows, draw down the cutting edge of the blade until the bevel is about 10 mm long.

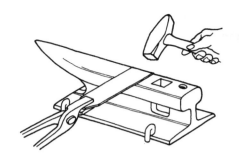

Anneal the plane iron and file to shape. All surfaces must be filed until a smooth flat finish is obtained.

The blade can now be hardened and tempered to straw colour in exactly the same way as the 25 mm chisel (see page 114).

The finished plane iron can now be ground to the correct angle and sharpened.

10 mm mortise chisel

The 10 mm mortise chisel is a very useful tool for the carpenter. This size of chisel is commonly used for cutting out the mortises in tables and chairs.

The chisel is used in the same way as a factory-made chisel except that the handle is struck with a hammer rather than a mallet.

The finished chisel should look like this.

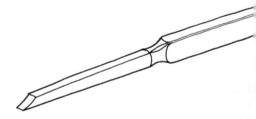

The drawings show the dimensions of the finished tool.

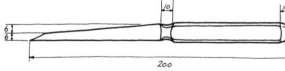

Cut off and straighten out 180 mm of 14–16 mm diameter coil spring (see pages 23–4).

Make a groove in the centre of the workpiece using the simple top and bottom fullers described on page 47. The fullering should be done in three stages. First, a groove is made in the top and bottom of the metal. It is then rotated a quarter of a turn and the metal fullered into a square section. Still working with the top and bottom fullers, the corners can finally be removed as shown in the drawings.

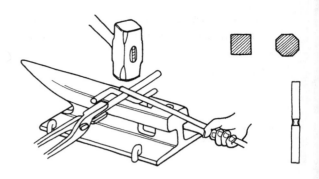

With half of the chisel heated up to its normal working temperature, flatten the handle.

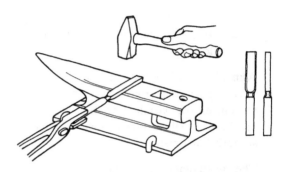

Form the striking end of the handle by chamfering the end and squaring up.

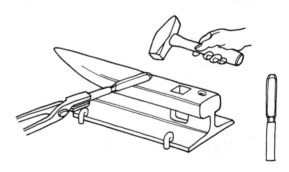

Draw down the blade so that the four sides are straight and the width is even along its length.

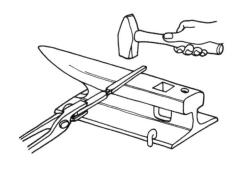

Cut the blade to length (100 mm) using a hot set and forge the cutting edge until the bevel is 12 mm long.

Anneal the chisel and carefully file into shape. It is important to make sure that the sides of the chisel are parallel for the whole length of the blade. All surfaces must be filed until a smooth flat finish is obtained.

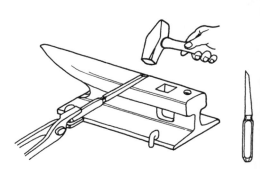

The chisel blade can be hardened and tempered to a pale straw colour either by the one-heat method described in the section on cold chisels (see pages 30–31), or in the same way as the 25 mm chisel (see page 114).

The blade can now be ground to the correct angle and sharpened.

Carpenter's brace

The factory-made carpenter's brace is usually a very expensive tool to buy. The chuck which holds the bit is very delicate and, if it breaks, it is difficult or sometimes impossible to repair.

This section will show you how to make a simple but strong alternative. The finished tool should look like this.

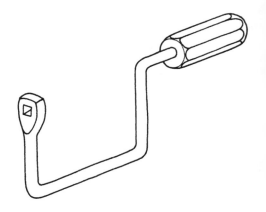

The two drawings here show the dimensions of the finished tool.

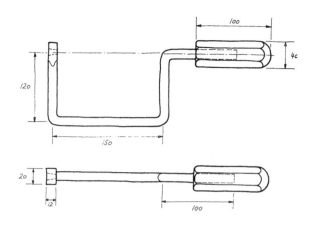

To make the tool you will need a square punch. To make one you can use the techniques described on pages 23–4. The size and shape of the tip of the punch should be carefully copied from the tapered shank of an old wood drilling bit. Use hot filing techniques to finish the punch.

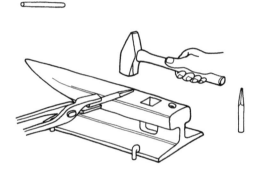

Starting with a 600 mm length of 12 mm diameter round mild steel bar, start upsetting the end. This end will eventually form the chuck. Use water to cool the parts of the bar which you do not want to upset.

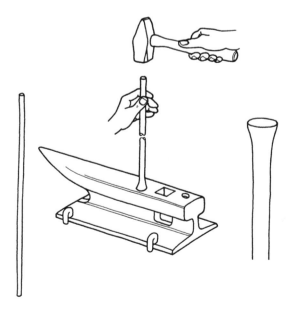

119

Chamfer the end of the bar as shown.

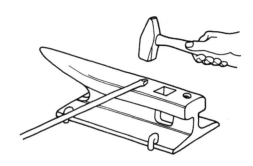

Continue to upset the bar until the diameter at the tip has reached 20 mm.

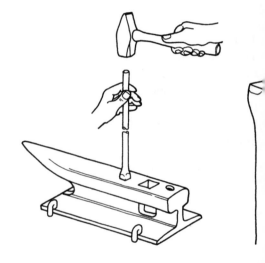

Flatten the upset part of the bar until its thickness is equal to the diameter of the bar.

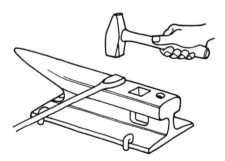

The chuck is now ready to have the hole punched in it.

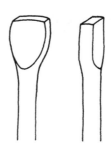

Using a square punch, make a tapered square hole in the centre of the flattened area. To do this, punch the metal most of the way through from one side, working on the flat part of the anvil.

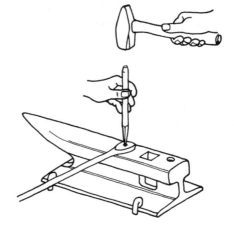

Turn the metal over and, working over the pritchel hole, drive the punch in just far enough to form a small hole through the chuck. Turn the metal back over and use the punch to open up the hole to the correct size.

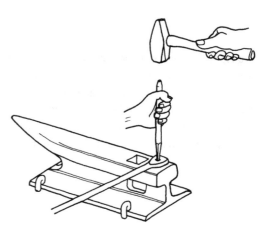

The finished chuck should look like this.

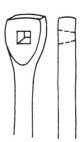

Mark two chalk lines to show the position of the first two bends as shown below.

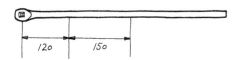

Use a hand hammer and the edge of the anvil to make a small dent at the position of each bend.

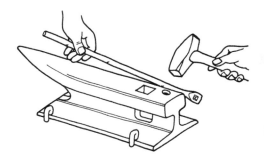

Next, the bends can be made as shown. When you are making the first bend, the larger side of the hole in the chuck must be facing away from you. Before each bend is made, cool down the metal on either side of the dent so that it is only hot around the area to be bent.

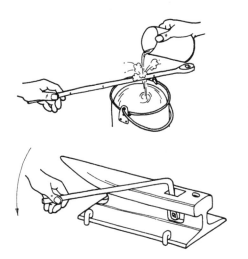

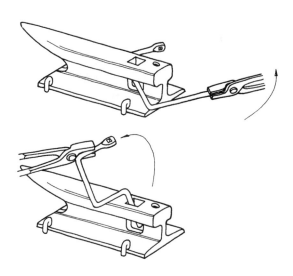

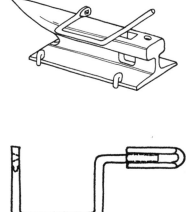

The position of the last bend can then be marked. The brace is laid along the anvil as shown, with the hole in the chuck level with one edge of the anvil and the handle lying parallel to the other edge. A chalk mark should be made at the point where the metal crosses the far edge of the anvil. The final bend can be put in the handle at this point.

Finally, check that the hole in the chuck is in line with the top part of the brace and adjust as necessary. The final adjustment can be done cold. The brace should now be fitted with a simple hardwood handle as shown.

20 mm centre bit

A simple but effective drill bit can be made from a short piece of straightened-out coil spring. When finished, it should look like this.

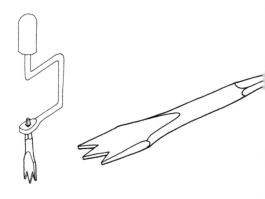

The two drawings show the shape and dimensions of the finished tool.

Using the techniques described earlier in the book (see pages 23–4) cut off and straighten out a 100 mm length of 12 mm diameter coil spring.

Carefully draw down a tapered square to fit the brace made earlier. Remove any sharp edges by hot filing. The length of the taper should be around 40 mm.

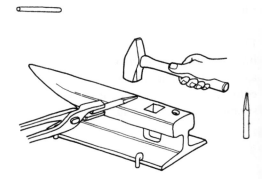

Heat up the other end of the bit and draw it down to a wedge-shaped taper.

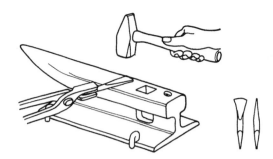

Forge in the sides of the taper until they are parallel.

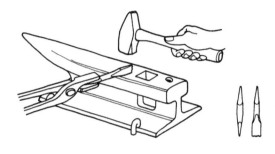

Using a thin hot chisel carefully cut a **V** in the end of the bit as shown. Remember to protect your anvil with a piece of scrap mild steel sheet.

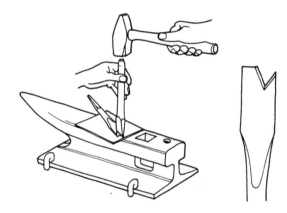

Turn the bit over and cut a **V** in the other side.

Anneal the bit and file to shape. Make sure that the tip is exactly in the centre of the bit and that the cutting blades have been filed to give a small clearance when in use.

The bit can now be hardened and tempered to a dark blue colour using the same techniques as used for the 25 mm chisel (see page 114).

QUESTIONNAIRE

So that Intermediate Technology can gain a greater understanding of the needs of blacksmiths throughout the world, we would like to know how this manual is used and what you think of the contents. When you have made a few of the tools, please answer as many of the questions as you can and send this page to IT Publications, 103–105 Southampton Row, London WC1B 4HH, UK.

Your work

1. COUNTRY _____

2. TYPE OF WORK _____ _____

3. PLACE OF WORK
    - ☐ Secondary school   ☐ Government training centre
    - ☐ NGO training centre   ☐ Other _____

The manual

4. How did you find out about the manual?
    - ☐ Through Intermediate Technology   ☐ Through VSO
    - ☐ Local library   ☐ Other sources _____

5. Did you find the manual easy to understand? If no, please give reasons:
    _____
    _____

6. Did you find the information in the manual useful? Please give reasons:
    _____
    _____

The tools

7. Have you made any of the tools described in this manual? If yes, which ones?
    - ☐ Round punch   ☐ Fullers
    - ☐ Hot chisel   ☐ Hammer
    - ☐ Cold chisel   ☐ Axe
    - ☐ Hot set   ☐ Hoe
    - ☐ Cold set   ☐ Knife
    - ☐ Tongs   ☐ Sickle

8. Have you had any problems with tools you made? If yes, give reasons:
    _____
    _____

Please use the back of the questionnaire for further comments.

www.ingramcontent.com/pod-product-compliance
Lightning Source LLC
Jackson TN
JSHW062202130125
77033JS00018B/601